# BASIC CHEMISTRY

## A programmed presentation

**Stewart M. Brooks**

Chairman of Science Curriculum
Newton-Wellesley Hospital School of Nursing
Newton Lower Falls, Massachusetts

**Cynthia F. Norton**

Department of Biology
University of Maine at Augusta
Augusta, Maine

**Third edition**

**Saint Louis**

**The C. V. Mosby Company**

**1976**

**Third edition**



Printed in the United States of America

Distributed in Great Britain by Henry Kimpton, London

**Library of Congress Cataloging in Publication Data**

Brooks, Stewart M
Basic chemistry.

Bibliography: p.
1. Chemistry—Programmed instruction.
I. Norton, Cynthia F., 1940- joint author.
II. Title.
QD31.2.B77 1976 540'.7'7 75-37515
ISBN 0-8016-0797-3

TS/VH/VH 9 8 7 6 5 4 3 2 1

# PREFACE

This learning program has been tailored to the level of a first course in chemistry for biologically oriented curricula. Experience with the first two editions shows that with proper use it can be helpful in many ways, perhaps above all as a reassuring companion through those proverbial trouble spots. Many students use the program as a review of a course once taken but much forgotten; such review can make a great deal of difference in the study of biology in general and metabolic systems in particular.

Dual authorship has obvious and considerable advantages. In this text, each and every frame was scrutinized by us together, and the final product is commensurate with our combined wishes: a clear, concise, complete, and correct—in a word, helpful—program.

**Stewart M. Brooks**
**Cynthia F. Norton**

# INSTRUCTIONS

## For the use of this program

In this programmed text, a question and its answer constitute a *frame*. The answers to each question are in the narrow left-hand margin of each page.

The most fruitful way to use programmed material is to mask the answer column with a sheet of paper and then work through the questions in order. Read a question carefully, consider its answer, and write the answer in the space provided. Then uncover the answer to that question by sliding the paper mask down the page to a point even with the bottom of the question. If your answer is correct, move on to the next question in the same manner. If your answer is incorrect, be sure that you understand the correct answer before reading the next question. Sometimes you may need to go back a few frames or even to the beginning of the section to review and master a concept before proceeding.

Getting the most out of programmed learning requires self-discipline. It is very easy to cheat or to simply read the questions and the answers. But a concept is most efficiently learned through your active participation in answering questions yourself and using the answers given only to check your understanding. If you cheat, you only fool yourself into thinking that you understand.

# CONTENTS

## PART IV
## Descriptive chemistry of some common compounds



## PART V
## Stoichiometry



## PART VI
## Gas laws



## PART VII
## Solutions



## PART VIII
## Colloids

## PART IX
## Nuclear chemistry



## PART X
## Organic chemistry: the hydrocarbons



## PART XI
## Organic chemistry: substituted hydrocarbons



## PART XII
## Biochemistry



## PART XIII
## Metabolism

# BASIC CHEMISTRY

## A programmed presentation

**PART I**

# The physical universe

## Section 1: MATTER

**1** The universe as we know it is composed of various types of matter disposed in space. Matter is acted upon by energy, resulting in motions and changes within the matter. The chemist is primarily involved with studying matter, the physicist with studying the effects of energy on matter.

**2** Matter is defined as anything that occupies space and has *mass*. For our purposes, we will treat *mass* and *weight* as the same thing. An object such as a chair, which occupies space and has mass, is made
matter — up of ________________.

**3** Under ordinary conditions, we apply the Law of Conservation of Mass, which states that mass can neither be created nor destroyed. When we carry out a chemical experiment and at the end observe exactly the same amount of matter we started with, we are illustrating
Conservation of Mass — the Law of ________________ of ________________.

**4**
create — In other words, in the laboratory we cannot ________________
matter — new matter or destroy existing ________________ under ordinary conditions.

## Section 2: PHYSICAL CHANGES IN MATTER

**1** Although mass (matter) cannot normally be created or destroyed, it can undergo many sorts of changes. Physical changes result in alteration of a physical property, without a new type of matter being
will not — formed. A physical change in matter then *will/will not* result in a totally new type of matter.

**2** The most important physical changes are those that result in a change of the *state* of matter. Matter of any type may exist in three states: solid, liquid, and gas. Which state is present at any given time depends on the temperature of the matter. The common type of matter that we call water, for example, can be observed in all three states:
liquid — solid, ________________, and gas.

**3**
solid — Ice, the ________________ state of water, is found at low temperatures.

water — **4** Steam, the gaseous state of ___water___, is found at
high — ___high___ temperatures.

**5** A solid has a *fixed* shape and occupies a *fixed* volume, or amount,
water — of space, as does a block of ice. A liquid, such as ___water___, has a *fixed* volume but a *variable* shape that changes to fit any container.

**6** A gas, such as steam, has both a *variable* shape and a *variable*
volume — ___volume___, expanding to fill the space available.

fixed — **7** A solid, then, has both shape and volume ________.

fixed — **8** A liquid, on the other hand, has a ________ volume but
shape — a variable ________.

**9** Because water may be changed back and forth from one state to another any number of times without ceasing to be water, we see
physical — that change of state is a clear example of a ________ change in matter.

## Section 3: CHEMICAL CHANGES IN MATTER

**1** Many types of changes in matter result in the appearance of a completely different type of matter. These are chemical changes. When paper is burned, ashes and a mixture of gases and smoke are formed. These are new types of matter, not paper in a different form. There-
chemical — fore, burning paper results in a ________ change.

**2** Chemical changes are usually referred to as *reactions*. The rusting of
reaction — iron is another common example of a chemical ________.

**3** Chemical reactions, as opposed to physical changes, may not be readily reversible. For example, ashes and smoke cannot be readily recombined to form paper. On the basis of what you have learned, classify each of the following as a physical or chemical change:

physical — Heating butter to melt it ________
chemical — Combustion of gasoline in an auto ________
physical — Casting a bronze statue ________
chemical — Developing a photographic negative ________

## Section 4: CHEMICAL CLASSIFICATION OF MATTER

**1** Matter can be separated into *elements, compounds,* and *mixtures*. Elements are the simplest of these, being made up of only one type of material. Thus, we speak of pure iron or pure oxygen gas as
elements — ________. All of the elements are shown in the table at the front of this book. Some common elements and their chemical *symbols* are listed in Table 1.

**2** Familiarize yourself with the symbols for each of these common elements. You will be using them frequently. For example, the ma-
potassium — terial with the formula $K_2CO_3$ contains ________,
carbon — ________, and oxygen.

**Table 1.** Common elements

| *Element* | *Symbol* | *Element* | *Symbol* |
|---|---|---|---|
| Aluminum | Al | Magnesium | Mg |
| Barium | Ba | Manganese | Mn |
| Boron | B | Mercury | Hg |
| Bromine | Br | Nickel | Ni |
| Calcium | Ca | Nitrogen | N |
| Carbon | C | Oxygen | O |
| Chlorine | Cl | Phosphorus | P |
| Chromium | Cr | Platinum | Pt |
| Cobalt | Co | Potassium | K |
| Copper | Cu | Silicon | Si |
| Fluorine | F | Silver | Ag |
| Gold | Au | Sodium | Na |
| Helium | He | Sulfur | S |
| Hydrogen | H | Tin | Sn |
| Iodine | I | Uranium | U |
| Iron | Fe | Zinc | Zn |
| Lead | Pb | | |

**3** Give the symbols for each of the following:

| | | |
|---|---|---|
| Cl | Chlorine | ________ |
| Au | Gold | ________ |
| Mg | Magnesium | ________ |
| P | Phosphorus | ________ |
| Sn | Tin | ________ |

compound

**4** A *compound* is formed when two or more specific elements combine by a chemical reaction. A new material is thus formed. The participating elements lose their identity, and the newly formed ________ has its own characteristic properties.

elements

**5** When a compound is formed from two or more ________________, the Law of Definite Proportions applies. This law states that the participating elements combine in *fixed proportions* by weight. Means it'll always be such as long as it's that compound

hydrogen

Definite Proportions

**6** Thus water, which has the commonly seen formula $H_2O$, is a compound. It contains two elements, ________________ and oxygen, in a fixed ratio of 2 particles (atoms) of hydrogen to 1 of oxygen. Water, therefore, is a true compound and fulfills the Law of ________________ ________________.

mixtures

**7** Substances that are neither elements nor compounds are mixtures. A mixture is any combination of elements and compounds in *variable proportions*. Many common materials such as soil, milk, blood, and air are ________________. The constituents of a mixture are usually easily separated, for example, gravel from soil or butterfat from milk. On the basis of what you have learned, classify each of the following as element, compound, or mixture:

| | | |
|---|---|---|
| mixture | Housepaint | ________________ |
| element | Nitrogen gas | ________________ |
| compound | Carbon dioxide | ________________ |
| element | Copper metal | ________________ |
| mixture | Orange juice | ________________ |

## Section 5: DENSITY

**1** Density is a physical property of matter. A term commonly encountered in chemistry, *density* is defined as the weight per unit volume of a given substance. It is computed by dividing the ________________ of a certain amount of a substance by its volume.

weight

**2** The density of liquids and solids is usually expressed in grams per milliliter (g/ml) or grams per cubic centimeter (g/cc), and that of gases in grams per liter (g/l). Thus, if we wish to calculate the density of a certain liquid, and we know that 20 ml of it weighs 40 g, we divide the weight, ____________, by the volume, ____________.

40 g; 20 ml

**3** The density of this liquid, then, is ______ g/ml.

2

**4** The density of a liquid of which 36 ml weighs 54 g is ________.

1.5 g/ml

**5** To measure the volume of a solid object accurately, we use the principle discovered by Archimedes. Suppose a piece of metal is dropped into a graduated cylinder containing water and the water level rises from the 40-ml to the 55-ml mark. That piece of metal has displaced 55 − 40, or 15 ml of water, and therefore has a volume equal to 15 ml. If a different piece of metal causes the water level to rise from the 42-ml to the 77-ml mark, it has a volume of ______ ml.

35

**6** Suppose we weighed this second piece of metal before wetting it and found it to weigh 60 g. We can now calculate its density as ________________.

1.7 g/ml

## Section 6: ENERGY

**1** Energy is customarily defined as the capacity to do work. In contrast to matter, it does not have mass and does not occupy ____________.

space

**2** Although the study of energy is primarily the province of the physicist, we must be concerned with it too in our study of chemistry. Every change in matter, whether physical or chemical, is accompanied by or caused by the use of energy. Whenever some form of work has been done, we know that ________________ has been used.

energy

**3** Under ordinary circumstances, we apply the Law of Conservation of Energy, which states that energy may neither be created nor destroyed. When we carry out a process involving the use of energy, we will have the same amount of energy at the end of the process as at the beginning. This observation illustrates the Law of ________________ of ________________.

Conservation of Energy

**4** Energy may exist in many forms. One of these is *mechanical* energy, the energy associated with movement. Falling water, moving autos, and sound are all forms of ________________ energy.

mechanical

**5** *Chemical* energy is the energy used to manufacture chemical compounds from ________________. It is released when compounds are taken apart.

elements

**6** *Electromagnetic* energy includes visible light and invisible radiation such as ultraviolet, infrared, gamma, and x-rays.

**7** *Electrical* energy is the energy of a moving current of electric charges or a lightning flash.

**8** *Heat* energy is related to mechanical energy. When heat is applied to matter, it causes the particles to move faster. The effect is responsible
steam — for the change of water to ________, for example.

**9** The energy given off by a candle flame is predominantly light, or
electromagnetic; heat — ________ energy. ________ energy is given off also.

mechanical — **10** A rock rolling down a hill has ________ energy.

**11** Gasoline is a good fuel because its constituent chemical compounds
chemical — contain ________ energy.

## Section 7: TRANSFORMATION OF ENERGY

**1** In use, energy is constantly changed from one form to another. Let us first examine the most basic type of change that energy undergoes. Some energy is found in stored form, known as *potential* energy. A can of oil contains potential energy—it is not presently doing work, but when burned it will. A car's storage battery is a reservoir of
potential — ________ electrical energy.

**2** Energy that is actually in the process of doing work is called *kinetic*
kinetic — energy. The rock rolling down the hill is a body that has ________ mechanical energy.

**3** Potential energy and kinetic energy can be readily interconverted.
potential — For example, the ________ energy of a gallon of gasoline
kinetic — is readily converted into the ________ mechanical energy of a moving car.

**4** Conversions from one type of energy to another are also very common. For example, when we make a piece of toast, the toaster con-
electrical; heat — verts ________ energy into ________ energy to brown the bread.

**5** Describe the energy transformation that takes place in each of these devices:

| | | | | |
|---|---|---|---|---|
| electrical; electromagnetic (light) | Flashlight | ________ | ⟶ | ________ |
| electromagnetic; heat | Solar evaporator | ________ | ⟶ | ________ |
| electrical; mechanical | Stereo speaker | ________ | ⟶ | ________ |
| chemical; heat | Oil burner | ________ | ⟶ | ________ |
| electrical; mechanical | Electric toothbrush | ________ | ⟶ | ________ |

## Section 8: MEASUREMENT OF ENERGY

**1** Each form of energy may be measured and expressed in units. In chemistry we will be using primarily the units of heat measurement. For each type of energy, heat included, two types of units are employed: one to measure *quantity* and one to measure *intensity,* or effect of the energy applied.

heat

expands

**2** A thermometer measures the effect of ____________ energy in expanding a fluid in a tube. The greater the intensity of heat, the more the fluid, usually alcohol or mercury, ________________.

180

**3** The tube of a thermometer is divided into units called *degrees.* On the Fahrenheit thermometer used for everyday or household measurements, water freezes at 32° F and boils at 212° F. The scale between these points is divided into 212 − 32, or ________ units or degrees.

100

**4** On the Celsius, or centigrade, thermometer used in scientific measurement, water freezes at 0° C and boils at 100° C. The scale between these points is divided into ________ units.

smaller

**5** A degree Fahrenheit is thus a *larger/smaller* unit of heat than a degree Celsius.

1.8 (or 9/5)

**6** 1° C equals:

$$180/100 = \_\_\_\_\_\_\_\_ °F.$$

**7** To convert from one scale to another, we can use either of two formulas:

$$F = 1.8\ C + 32$$

$$C = \frac{F - 32}{1.8}$$

~~50.8~~ 60.8
13.3
−11.1
230

**8** Convert the following:

16° C = ________ °F
56° F = ________ °C
12° F = ________ °C
110° C = ________ °F

quantity

**9** The quantity of heat energy is usually expressed in *calories* (cal). The calorie is defined as the ________________ of heat necessary to raise the temperature of 1 g of water 1° C.

105

**10** The quantity of heat required to raise the temperature of 35 g of water 3° C is ________ calories.

45

**11** Food energy is usually expressed in kilocalories, or large calories (Cal). One Cal equals 1,000 cal. A quantity of the food in question is burned, and the heat released is measured in a calorimeter. Most fats, for example, have 9 Cal per gram. If 5 g of fat were burned, ________ Cal of heat would be liberated.

2

**12** If 12 g of an unknown substance were burned and the temperature of 1,000 g of water was raised 24 degrees, the substance would have an energy content of ________ Cal per gram.

PART **II**

# Structure of matter

## Section 1: THE ATOM

**1** All forms of matter are composed of very small basic particles called *atoms*. In this section we shall discuss the structure of the atoms of different elements. Atomic structure controls the types of activities or reactions that elements undergo and the compounds they form. The atom is the smallest unit of an element. Physicists, in studying the normally stable atom, have shown that it can be subdivided into an ever-growing number of *subatomic particles*. In our study of the atom we will concentrate on only three of these entities.

subatomic — **2** Each atom is made up of a fixed number of three ______________ particles: the *proton,* the *neutron,* and the *electron*.

**3** The proton has a mass (or weight) of 1 atomic mass unit (amu) and 1 unit of positive electrical charge. An atom that contains 10 protons,
10 — then, will have ______ positive electrical charges.

**4** The *neutron* likewise has a mass of 1 amu. Neutrons, however, have no electrical charge. An atom with 12 neutrons will acquire from
12; 0 — them ______ atomic mass units and ______ electrical charges.

**5** The *electron* has a negligible mass of 0.0005 amu. Each electron, however, has 1 unit of negative electrical charge. Therefore, an
0.01 — atom with 20 electrons will have ______ units of mass and a total
20 — of ______ negative charges.

**6** Protons and neutrons are found within the atom's dense central
positive — *nucleus,* which therefore has a ______________ electrical charge.

**7** Electrons are arranged around the nucleus in elliptical *orbitals*. The
electrons — relationship of the nucleus of an atom to its orbital ______________ can be compared loosely with the relationship of the sun to the planets in our solar system. The orbital electrons surround the nucleus with
negative — a number of ______________ electrical charges.

**8** The chemical behavior of any atom is determined by the number and arrangement of its electrons. We will defer further analysis
protons — of the nucleus and its particles, the ______________ and
neutrons — ______________, to Part IX.

## Section 2: PERIODIC TABLE

periodic **1** Refer to the periodic table at the front of this book. The ________ table of elements shows all known elements, arranged in an order corresponding to the number of protons, or positive charges, each contains. This is the *atomic number,* given in the upper lefthand corner of each box. For magnesium, then, the atomic number is
12 ______.

10 **2** Neon can be seen to have ______ protons.

**3** The number at the bottom of the box, the *atomic weight,* gives the average mass of an atom of the element. Thus, the average mass of
35.453 an atom of chlorine is __________.

**4** Most atomic weights are not whole numbers, because atoms of these elements exist in two or more different forms called *isotopes.* Isotopes of an element differ from one another only in the number of *neutrons* they possess.

**5** The atomic weight of an element will be an average of the masses of the respective isotopes, each of which will have a different mass be-
neutrons cause it has different numbers of ______________.

**6** The next two sections show why the elements are arranged as they are in the periodic table.

## Section 3: ELECTRON ORBITALS

**1** Electron orbitals represent mathematical descriptions of the probable location of an atom's electrons revolving around the nucleus. The
orbitals study of electron ______________ is called quantum mechanics.

**2** Electrons are found at quantum levels representing average distances from the nucleus. The *primary quantum number* for an electron is *n*. The closest electrons are found at the quantum level where n = 1.
closer to There are 7 levels in all. An electron for which n = 3 is *closer to/ farther from* the nucleus than one at n = 5.

primary quantum number **3** The __________ __________ __________ (n) of an electron thus defines its probable distance from the nucleus.

**4** Each quantum level has a fixed capacity for electrons, as seen in
18 Table 2. Thus, n = 3 has a capacity for a maximum of ______ electrons.

**5** Each quantum level can be further subdivided into orbitals. Refer again to Table 2. The larger n is, the more complex is its substructure, up to a point. For our purposes, it is sufficient to learn which orbitals are found at each quantum level.

**6** A quantum level will have specified numbers of different types of orbitals: n = 2 will have 1 s and 3 p orbitals. Each orbital contains a
8 maximum of 2 electrons; thus n = 2 will contain a total of ______ electrons.

**7** Because level n = 5 has s, p, d, and f orbitals, it can hold a maximum
32 of ______ electrons.

**Table 2.** Electron arrangements in atoms

| *Quantum level (n)* | *Orbitals* | *Maximum electron number* |
|---|---|---|
| 1 | 1 s | 2 |
| 2 | 1 s, 3 p | 8 |
| 3 | 1 s, 3 p, 5 d | 18 |
| 4 | 1 s, 3 p, 5 d, 7 f | 32 |
| 5 | 1 s, 3 p, 5 d, 7 f | 32 |
| 6 | 1 s, 3 p, 5 d | 18 |
| 7 | 1 s | 2 |

**8** The rules for the order in which orbitals are filled as atoms increase in size are complex. They are summarized in Fig. 1. Comparison of this figure with the periodic table is necessary.

**9** You can see that all elements in groups IA and IIA have one thing in
s — common: they are all elements in which the ______ orbitals are
p — being filled. Groups IIIA through VIIA have the ______ orbitals
being filled.

d — **10** The so-called transition elements have their ______ orbitals being filled.

**11** The periodic table was arranged before anyone suspected the existence of electrons. It was arranged on the basis of the atomic weight of known elements, so that elements with similar chemical behavior would be in groups, as represented by the numbered vertical columns. By looking at the table we can say that sodium will resemble
more — potassium *more/less* than silicon.

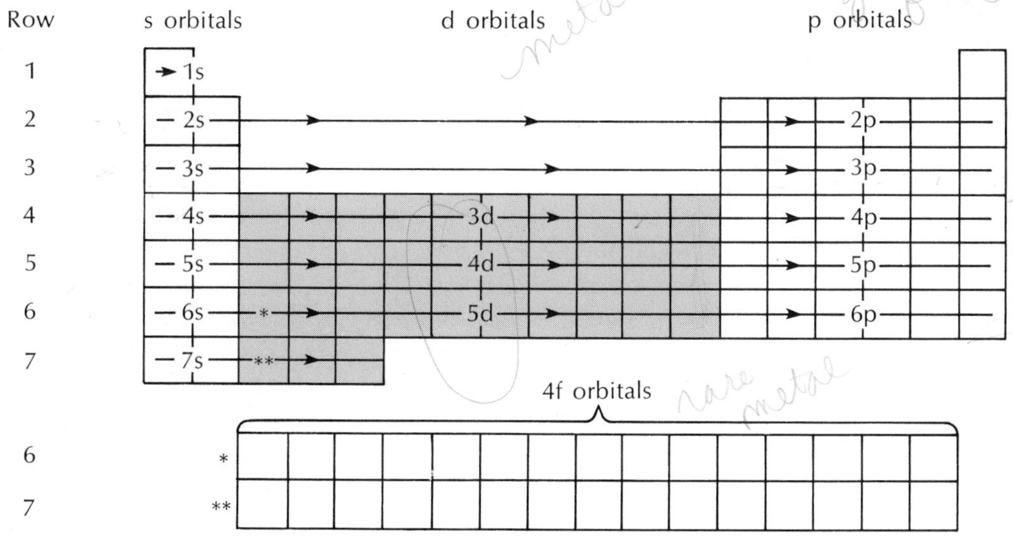


**Fig. 1.** Relationship of the increase in the number of electrons to the arrangement of elements in the periodic table. Refer to periodic table inside front cover. (From Drago, R. S.: Principles of chemistry with practical perspectives, Boston, 1974, Allyn & Bacon, Inc.

phosphorus

**12** Nitrogen's chemical behavior most closely corresponds to *phosphorus/oxygen.*

p

**13** With our knowledge of orbital structure we can see why fluorine and chlorine are chemically similar: they are missing just 1 electron from their ______ orbitals.

## Section 4: PROPERTIES OF ELEMENTS RELATED TO ATOMIC STRUCTURE: VALENCE

2

**1** For maximum stability, atoms need an *outer* quantum level containing either 2 electrons for $n = 1$ or 8 electrons for the others. Chemical stability for $n = 2$ is obtained when the s and p orbitals each contain ______ electrons, for a total of 8.

neon

**2** An element corresponding to this description is ______ (see periodic table).

1

+2

**3** Atoms that do not have the stable numbers of electrons in the outer quantum level undergo chemical reactions to gain, lose, or share electrons. The *valence* of an atom is the number of electrons it must gain, lose, or share to reach stability. For example, sodium needs to lose 1 electron; thus it has a valence of ______. Because it now has 1 more proton than electrons, the valence is given a positive (+) sign, that is, +1. Magnesium, with 2 electrons to lose, has a valence of ______.

−1

**4** Fluorine, which needs 1 more electron to reach stability (has 1 to gain) has a valence of ______.

quantum

**5** Elements with positive valences (Groups IA, IIA, and the transition elements) are metals. The other elements (Groups IIIA through VIIA) are nonmentals. The so-called *noble gases* (Group O) do not have a valence because they all have full outer ______ levels.

metal; noble gas; nonmetal

**6** Thus strontium is a ______, argon is a ______ ______, and sulfur is a ______.

gives

0

8

**7** *Electrovalence* is the term used when one atom gives electrons to another. For example, sodium (valence +1) reacts with chlorine (valence −1) to form the compound sodium chloride. In this reaction a sodium atom ______ 1 electron to a chlorine atom. Sodium now has a valence of ______, as does chlorine. The outermost quantum level of each atom now has ______ electrons.

electrovalent

ion

**8** Sodium has lost 1 electron in this electrovalent reaction. After the reaction, it has 1 more proton than it has electrons; therefore, it has a net electrical charge of +1 (the same expression as for its valence). Atoms that have acquired an electrical charge as a result of an ______ reaction are called *ions*. Sodium is an example of a positively charged ______. Such ions are called *cations*.

−1

**9** Chlorine has gained 1 extra electron; thus it has a charge of ______. Negatively charged ions, such as $Cl^-$, are called *anions*.

lose; cations

**10** Atoms with positive valences *gain/lose* electrons and form *cations/anions*.

gain **11** An element with a valence of −2 will __________ 2 electrons in
anion an electrovalent reaction and form a negative ion, or __________,
−2 with an electrical charge of ______.

**12** Chemically combined groups of atoms can participate in electrovalent reactions. They are called *polyatomic ions,* or radicals, and may be positively or negatively charged. A grouping such as am-
polyatomic monium ($NH_4^+$) is a positively charged __________ ion.
A positively charged polyatomic ion is a cation, and a negatively
anion charged one is a(n) __________.

**13** Atoms that share electrons undergo *covalent* reactions and form covalent bonds. One shared *pair* of electrons constitutes a *covalent bond*.

**14** Let us use water as an example of a compound formed by atoms shar-
covalent ing electrons and containing __________ bonds. In Fig. 2 a hydrogen atom is represented by the letter H designating its nucleus and a small x representing the single electron in its one and only quantum level. Oxygen is shown with the letter O representing its nucleus and six dots representing the 6 electrons of its second quantum level. The electrons of the first quantum level are not shown because they do not participate in the reaction. A hydrogen atom, then, has 1
1 electron in its outermost quantum level and must share ______ more
2 electron(s) to achieve stability. An oxygen atom must share ______
more electron(s) to become stable.

**15** Therefore, complete stability can be achieved if 1 oxygen atom combines with 2 hydrogen atoms. This gives us the familiar formula of
compound $H_2O$ for water. Water, a chemical __________, is made up of *molecules*. A molecule is the smallest unit of a *compound* and
definite consists of 2 or more atoms in __________ proportions.

molecule **16** A __________ of water, then, is made up of 2 atoms of
covalent hydrogen and 1 atom of oxygen, held together by 2 __________
bonds.

electrons **17** Each covalent bond consists of a pair of shared __________.
One of these is contributed by each of the participating atoms.

**18** Many elements can form several covalent bonds. For example, Table
4 3, on p. 12, shows that carbon always forms ______ covalent bonds. Because oxygen forms 2 bonds, we can predict that 1 carbon atom could react with 2 oxygen atoms if each shared two pairs of electrons. The Lewis dot structure for the resulting compound, $CO_2$, is shown in
two Fig. 3. The carbon atom shares __________ pair(s) of electrons with each oxygen atom. These are so-called *double bonds*. By comparison, because oxygen shares one pair of electrons with each of 2
single hydrogen atoms, water contains 2 __________ bonds.

**Fig. 2.** Molecular structure of water.

**Fig. 3.** Molecular structure of carbon dioxide.

**Table 3.** Valences of common elements and polyatomic ions

| *Element* | *Valence* | *Ions* |
|---|---|---|
| **Electrovalent** | | |
| *Metals* | | |
| Aluminum | 3 | $Al^{+++}$ |
| Barium | 2 | $Ba^{++}$ |
| Bismuth | 3 | $Bi^{+++}$ |
| | 5 | $Bi^{+++++}$ |
| Calcium | 2 | $Ca^{++}$ |
| Copper | 1 | $Cu^{+}$ |
| | 2 | $Cu^{++}$ |
| Hydrogen | 1 | $H^{+}$ |
| Iron | 2 | $Fe^{++}$ |
| | 3 | $Fe^{+++}$ |
| Lead | 2 | $Pb^{++}$ |
| Magnesium | 2 | $Mg^{++}$ |
| Mercury | 1 | $Hg^{+}$ |
| | 2 | $Hg^{++}$ |
| Nickel | 2 | $Ni^{++}$ |
| Potassium | 1 | $K^{+}$ |
| Silver | 1 | $Ag^{+}$ |
| Sodium | 1 | $Na^{+}$ |
| Tin | 2 | $Sn^{++}$ |
| | 4 | $Sn^{++++}$ |
| Zinc | 2 | $Zn^{++}$ |
| *Polyatomic ions* | | |
| Ammonium | 1 | $NH_4^{+}$ |
| Bicarbonate | 1 | $HCO_3^{-}$ |
| Carbonate | 2 | $CO_3^{=}$ |
| Chlorate | 1 | $ClO_3^{-}$ |
| Hydroxyl | 1 | $OH^{-}$ |
| Nitrate | 1 | $NO_3^{-}$ |
| Nitrite | 1 | $NO_2^{-}$ |
| Phosphate | 3 | $PO_4^{\equiv}$ |
| Sulfate | 2 | $SO_4^{=}$ |
| Sulfite | 2 | $SO_3^{=}$ |
| *Nonmetals* | | |
| Bromine | 1 | $Br^{-}$ |
| Chlorine | 1 | $Cl^{-}$ |
| Fluorine | 1 | $F^{-}$ |
| Iodine | 1 | $I^{-}$ |
| Oxygen | 2 | $O^{=}$ |
| Sulfur | 2 | $S^{=}$ |
| **Covalent** | | |
| Carbon | 4 | |
| Hydrogen | 1 | |
| Nitrogen | 1,2,3,4,5 | |
| Oxygen | 2 | |
| Phosphorus | 3,5 | |
| Sulfur | 2,4,6 | |

## Section 5: PROPERTIES OF ELEMENTS RELATED TO ATOMIC STRUCTURE: ELECTRONEGATIVITY

**1** *Electronegativity* is defined as the tendency of an atom to attract electrons to itself. This particularly affects the shared electrons in the covalent bond. The nucleus of each atom will exert a certain pull on the pair of electrons, one of which is its own and one of which belongs to the other participating atom. The stronger the attractive force, the greater the ________________ of the atom.

electronegativity

**2** From left to right in the periodic table, atoms grow *more* electronegative. For example, sodium is less electronegative than magnesium. Aluminum is *more/less* electronegative than phosphorus.

less

**3** In a bond between calcium and bromine, electrons would be more strongly attracted by ________________.

bromine

**4** Where two atoms forming a bond have electronegativities that are nearly equal, electrons are shared, and we call the bond a ________________ bond. Where the two atoms are quite unbalanced in their electron-attracting powers, electrons are transferred from one to the other, and an ________________, or ionic, bond is formed. There are all graduations between these extremes.

covalent

electrovalent

**5** An intermediate bond occurs when electrons are shared unequally. The pair of electrons is attracted more strongly to the *more/less* electronegative partner. Thus the molecule, although electrically neutral, will have an area of negative charge near the more ________________ partner and an area of positive charge near the other. This effect characterizes a *polar covalent* bond.

more

electronegative

**6** Covalent bonds, then, can be either *polar* or *nonpolar*. In molecules containing nonpolar covalent bonds, the electrons are shared *equally/unequally*.

equally

**7** In molecules containing polar covalent bonds, the electrons are shared ________________, and the surfaces of the molecules carry areas of ________________ charge.

unequally

electrical

**8** In covalence, the molecule is either polar (dipole) or ____________.

nonpolar

**9** We can summarize this treatment of chemical bonding by stating that bonds between atoms of approximately equal electronegativity will be ________________ covalent. If the electronegativities are moderately unequal, the bonds will be __________ __________. If the electronegativities are considerably different, the bond will be ________________, or ionic.

nonpolar

polar covalent

electrovalent

## Section 6: MOLECULES AND CHEMICAL FORMULAS

**1** We have defined a molecule as the smallest unit of a ____________. A molecule will contain whatever atoms the compound contains, in definite proportions.

compound

**2** To write the formula of a compound we must first find out which elements are present. For example, we already know that sodium chloride contains two elements, ______________ and ______________.

sodium
chlorine

**3** Next we need to know what proportions of the various elements can be expected. This is determined by the *valences* of the elements. For example, we can verify that sodium has a valence of ______, and chlorine a valence of ______.

+1
−1

**4** To complete the formula we need only to apply the rule that atoms (ions) of sodium times valence of sodium equals atoms (ions) of chloride times valence of chloride (signs must be opposite). Thus:

$$1 \text{ atom Na} \times +1 = 1 \text{ atom Cl} \times -1$$

This system works for all types of simple bonding.

**5** Thus, the *empirical formula* for sodium chloride is NaCl. This formula tells us that 1 molecule of sodium chloride contains 1 atom of sodium and 1 atom of ______________.

chlorine

**6** Carbon dioxide is another example. This compound contains carbon and oxygen. Note that the valence of carbon is ______, and that of oxygen is ______.

4
2

**7** Complete the following equation for carbon dioxide:

$$\underset{\text{atoms C}}{\_\_\_\_} \times \underset{\text{valence C}}{\_\_\_\_} = \underset{\text{atoms O}}{\_\_\_\_} \times \underset{\text{valence O}}{\_\_\_\_}$$

$1 \times 4 = 2 \times 2$

**8** We now see that this compound must be written to represent 1 carbon atom and 2 oxygen atoms, or $CO_2$. The subscript 2 following the symbol for oxygen indicates that there are 2 ______________ ______________ in the molecule.

oxygen
atoms

**9** Write formulas for compounds made up of:

| | |
|---|---|
| Potassium and oxygen | ________ |
| Magnesium and oxygen | ________ |
| Calcium and fluorine | ________ |
| Barium and iodine | ________ |
| Aluminum and bromine | ________ |
| Aluminum and oxygen | ________ |

$K_2O$
MgO
$CaF_2$
$BaI_2$
$AlBr_3$
$Al_2O_3$

**10** When there are two elements in a compound, the less electronegative one is written first, by convention. For example, a compound of sodium and iodine has the formula ____________, and a compound of chlorine and silver is *AgCl/ClAg*.

NaI
AgCl

**11** Some atoms have two or more possible valences. For example, sulfur has valences of 2, ______, and ______. Thus, when sulfur combines with oxygen, two different compounds are possible. Valence 4 sulfur forms ____________, sulfur dioxide; valence 6 sulfur forms ____________, sulfur trioxide. Note how the prefixes di- and tri- refer to the number of atoms of ______________ in these molecules.

4; 6
$SO_2$
$SO_3$
oxygen

**12** Metals with two valences form two compounds. For example, copper can be either valence +1 or ______. Copper (+1) reacts with

+2

CuCl

$CuCl_2$

chlorine to form ____________. Copper (+2) with chlorine forms ____________. CuCl is called cupr*ous* chloride; $CuCl_2$ is called cupr*ic* chloride.

higher

+3

**13** The *-ous* suffix denotes the lower valence state, and the *-ic* suffix the ________________ valence state. Of the 2 chlorides of iron, ferric chloride contains iron in the higher valence state, ________.

$Fe^{+++}$

**14** Ferrous chloride contains the cation $Fe^{++}$, and ferric chloride contains the cation ____________.

+1

**15** If compounds contain polyatomic ions, these ions appear in the formula as a *unit*. Thus, ammonium, with its valence of ________ and formula of $NH_4$, is written directly into the formula for ammonium chloride as $NH_4Cl$.

−2

$BaSO_4$

**16** Sulfate ($SO_4^{=}$) has a valence of ________. Therefore, the formula for barium sulfate is ____________.

**17** When a polyatomic ion is present two or more times in a formula, its symbol is written in parentheses, with a subscript number outside to indicate the multiple. Thus the formula for aluminum hydroxide is $Al(OH)_3$.

**18** Write the formulas for the following compounds:

| | | |
|---|---|---|
| $CaSO_3$ | Calcium sulfite | ____________ |
| $Mg(OH)_2$ | Magnesium hydroxide | ____________ |
| $FeSO_4$ | Ferrous sulfate | ____________ |
| $Fe_2(SO_3)_3$ | Ferric sulfite | ____________ |
| $(NH_4)_3PO_4$ | Ammonium phosphate | ____________ |
| $KHCO_3$ | Potassium bicarbonate | ____________ |

## Section 7: WRITING EQUATIONS

created; destroyed

5

**1** In this section, the rules for writing chemical equations will be given. This is very simple if we remember the Law of Conservation of Mass: mass is neither ________________ nor ________________. For example, if we start out with 5 atoms of oxygen, we must end up with ________ atoms of oxygen.

**2** When a solution of silver nitrate is added to a solution of calcium chloride, a chemical reaction occurs that forms calcium nitrate and silver chloride. The latter is insoluble in water and settles out as a white precipitate. To express this reaction in the form of an equation, put the formulas for the starting materials, or *reactants,* to the left of an arrow and the *products* to the right:

$$CaCl_2 + AgNO_3 \longrightarrow Ca(NO_3)_2 + AgCl\downarrow \text{ (precipitates)} \qquad \text{(II-1)}$$

1

Conservation of Mass

**3** Note that we start out with 2 chloride ions and end up with ________. This *does not* obey the Law of ________________ of ________.

**4** This is an unbalanced equation, but can be corrected by the use of *coefficients*. Thus:

$$CaCl_2 + 2AgNO_3 \longrightarrow Ca(NO_3)_2 + 2AgCl\downarrow \qquad \text{(II-2)}$$

Checking this through, we see that there are now as many atoms on the left as there are on the right. For example, there are 2 atoms of nitrogen on each side, ______ atoms of oxygen on each side, and so forth. More conveniently, we say that there are ______ nitrates on each side, because the polyatomic ion behaves as a unit.

6

2

**5** Equation II-2 reads: *1* molecule of $CaCl_2$ (1 is understood) plus 2 molecules of silver nitrate *yields* ______ molecule(s) of $Ca(NO_3)_2$ and ______ ______________ of AgCl. The symbol ↓ indicates a(n) ______________.

1

2 molecules

precipitate

**6** Always reduce coefficients to the *smallest whole numbers*. Thus the equation

$$4\ Ag + 2\ S \longrightarrow 2\ Ag_2S \quad \text{(II-3)}$$

is balanced, but should be written

$$2\ Ag + S \longrightarrow Ag_2S \quad \text{(II-4)}$$

**7** Balance these equations (indicating 1 where no coefficient is needed):

$$\_\_\ HCl + \_\_\ AgNO_3 \longrightarrow \_\_\ AgCl + \_\_\ HNO_3 \quad \text{(II-5)}$$

1, 1, 1, 1

$$\_\_\ MgCl_2 + \_\_\ NaNO_3 \longrightarrow \_\_\ Mg(NO_3)_2 + \_\_\ NaCl \quad \text{(II-6)}$$

1, 2, 1, 2

$$\_\_\ KHCO_3 + \_\_\ AlBr_3 \longrightarrow \_\_\ Al(HCO_3)_3 + \_\_\ KBr \quad \text{(II-7)}$$

3, 1, 1, 3

$$\_\_\ (NH_4)_3PO_4 + \_\_\ Fe_2(SO_4)_3 \longrightarrow \_\_\ FePO_4 + \_\_\ (NH_4)_2SO_4 \quad \text{(II-8)}$$

2, 1, 2, 3

PART **III**

# Chemical reactions

## Section 1: DISSOCIATION REACTIONS

**1** When an electrovalent compound is dissolved in water, the ions composing the molecules of the compound may be driven apart. The ions then exist as free charged particles in solution. This process is referred to as *dissociation* (or, less correctly, as *ionization*). A dissociation reaction occurs when a(n) ________ compound is dissolved in ________. This type of reaction results in the release of free ________.

electrovalent
water
ions

**2** A solution of an electrovalent compound is electrically neutral because it contains an equal number of positive and negative *charges*. Because such compounds (in solution) conduct an electrical current, they are commonly referred to as *electrolytes*. The battery acid in a car's storage battery is a good example of an ________.

electrolyte

**3** Nonelectrolytes are compounds that do not ________ in water.

dissociate

**4** The amount of current passing through a solution indicates the extent of ________, that is, the number of dissociated molecules or free ________.

dissociation
ions

**5** The ability of a solution to pass current is called its conductivity. The greater the conductivity of a solution, the *higher/lower* the number of free ions present.

higher

**6** Some compounds, such as NaCl, undergo *complete* dissociation when dissolved in water. These are *strong* electrolytes and therefore *good/bad* conductors.

good

**7** Dissociation is expressed as follows:

$$NaCl \longrightarrow Na^+ + Cl^- \quad \text{(III-1)}$$

This reaction reads: 1 ________ of sodium chloride yields 1 sodium ion and 1 ________ ________.

molecule
chloride ion

**8** Compounds made up of atoms of widely different electronegativities often undergo complete ________. These are true ________ compounds.

dissociation
electrovalent

**9** *Nonpolar* covalent compounds do not undergo ________ when placed in water, although they may dissolve. Their solutions are therefore not ________ of electricity.

dissociation
conductors

electrolytes

**10** *Polar* covalent compounds undergo *partial* dissociation. Such compounds are thus relatively *weak* ________, or poor conductors.

NH₄OH / ammonium / conductivity

**11** To express the partial dissociation of a weak electrolyte such as $NH_4OH$, we write the equation as follows:

$$NH_4OH \rightleftharpoons NH_4^+ + OH^- \qquad \text{(III-2)}$$

The double arrow indicates that the reaction is *reversible*. This can be explained by saying that, although each of the many molecules of ________ we add to the solution may undergo dissociation, at the same time 1 ________ ion and 1 hydroxide ion may reassociate. The reaction does not progress either to the left or right, and only part of the ions will ever be in the free state. This condition results in poor electrical ________.

reassociation

**12** Partial dissociation results in an *equilibrium* state. The concept of equilibrium is very important in chemistry. Simply put, an equilibrium results when, as in the example in frame 11, the forward and back reactions progress at the same rate, or speed. Thus, dissociation and ________ are in equilibrium, or balance.

strong (good) / all / dissociate / poor (weak); few

**13** To summarize, then, electrovalent compounds are ________ electrolytes because *all/few/none* of their ions are free in aqueous solution. Nonpolar covalent compounds are nonelectrolytes because they do not ________ at all. Polar covalent compounds may be ________ electrolytes because *all/few/none* of their ions are free in solution at any one moment.

## Section 2: OXIDATION-REDUCTION REACTIONS

**1** *Electrolysis* is a process whereby a direct current of electricity passing through a solution of an electrolyte causes separation of the ions.

copper (cupric) / chloride

**2** Let us use a solution of $CuCl_2$ for discussion. $CuCl_2$ dissociates as follows:

$$CuCl_2 \longrightarrow Cu^{++} + 2Cl^- \qquad \text{(III-3)}$$

The electrodes supplying the current are immersed in a solution containing positive ________ ions and negative ________ ions.

**3** One electrode, the *anode*, has a positive electrical charge; the other, the *cathode*, has a negative charge.

cupric / cathode / chloride / anode

**4** Because of the rule that opposite electrical charges attract each other, we can predict that the positively charged ________ ions will migrate toward the negatively charged *anode/cathode*. Similarly, the negatively charged ________ ions will migrate toward the positively charged ________.

cathode / anions

**5** Positively charged ions are called cations, because they move toward the ________. Similarly, negative ions, or ________, migrate toward the anode.

**6** When the $Cu^{++}$ cation reaches the cathode, a layer of shiny metallic copper builds up on the cathode surface. The reaction that takes place is written as follows:

$$Cu^{++} + 2e^- \longrightarrow Cu^0 \qquad \text{(III-4)}$$

This equation shows that 1 cupric ion reacts with (gains) 2 electrons ($2e^-$) supplied by the cathode. This results in the formation of a copper atom that, like all atoms, is electrically neutral. The designation $Cu^0$ indicates electrical neutrality. To summarize, the cupric ion gains 2 __________ and becomes an electrically neutral __________ of copper.

electrons

atom

**7** The gain of electrons in a reaction is called *reduction*. Thus, in this reaction, copper is said to be *reduced*. Any reaction in which electrons are gained is called a __________ reaction.

reduction

**8** Now let us follow the chloride ions. These move toward the __________, and as a result bubbles of chlorine gas appear. Two reactions explain this event:

anode

$$Cl^- \longrightarrow Cl^0 + e^- \qquad \text{(III-5)}$$

$$2Cl^0 \longrightarrow Cl_2 \text{ gas} \qquad \text{(III-6)}$$

The first reaction shows that a chloride ion *loses* an electron to the positively charged anode. By doing so, it becomes a neutral chlorine atom $Cl^0$. The second reaction shows that 2 chlorine __________ combine to form a __________ of chlorine gas.

atoms

molecule

**9** The chloride ion has *gained/lost* electrons in the process. This is called *oxidation,* and the chloride ion is said to be *oxidized*. Thus, oxidation can be defined as a __________ of electrons.

lost

loss

**10** Combining the reactions involved in the electrolysis of $CuCl_2$, we have:

$$Cu^{++} + 2e^- + 2Cl^- \longrightarrow Cu^0 + Cl_2 + 2e^- \qquad \text{(III-7)}$$

*Both* oxidation and reduction have occurred. The reaction as a whole is called a *reduction-oxidation*-reaction, or *redox* reaction.

**11** If one substance *loses* electrons, or is __________, another must *gain* electrons, or be reduced. We cannot have one without the other in a redox reaction.

oxidized

**12** Let us look at a redox reaction that does not require batteries and electrodes. If we drop an iron nail into a solution of cupric sulfate, the nail rapidly becomes coated with copper metal. The ionic equation for this reaction is:

$$Fe^0 + Cu^{++} + SO_4^= \longrightarrow Fe^{++} + SO_4^= + Cu^0 \qquad \text{(III-8)}$$

This can be separated into two half-reactions, such as we used in our first example:

$$Fe^0 \longrightarrow Fe^{++} + 2e^- \qquad \text{(III-9)}$$

$$Cu^{++} + 2e^- \longrightarrow Cu^0 \qquad \text{(III-10)}$$

The sulfate ion ($SO_4^{=}$) has been left out because it does not undergo any changes during the reaction, but simply is capable of associating with whichever cation is present. The first equation shows a(n) ________________ reaction, whereas the second equation shows a(n) ________________ reaction.

oxidation
reduction

oxidized

**13** Iron is ________________ to ferrous ion by its interaction with cupric ion. Cupric ion, then, can be said to be an *oxidizing agent.* Similarily, cupric ion is reduced to copper by its interaction with iron; thus iron can be said to be a(n) ________________ agent in this reaction.

reducing

**14** Copper takes electrons away from, or oxidizes, iron, because it is more electronegative than iron. Table 4 shows the *electromotive series,* in which metals are arranged in order of their increasing oxidizing ability. Copper is lower down in the series than iron.

**15** In reaction III-8, simplified to

$$Fe^0 + CuSO_4 \longrightarrow FeSO_4 + Cu^0 \qquad \text{(III-11)}$$

iron replaces the ________________ ion.

cupric

**16** Table 4 also allows us to predict replacements. A metal higher in the series will always replace one lower in the series, as in the example of iron replacing copper. Thus zinc will replace *magnesium/cadmium.*

cadmium

**17** Hydrogen is included in the electromotive series because it acts like a metal; that is, it forms *positive/negative* ions.

positive

**18** Hydrogen *will/will not* take electrons from silver, but *will/will not* take electrons from sodium.

will not; will

**Table 4.** Electromotive series

| Increasing reducing ability ↑ | | Increasing oxidizing ability ↓ |
|---|---|---|
| | Lithium | |
| | Potassium | |
| | Sodium | |
| | Calcium | |
| | Magnesium | |
| | Aluminum | |
| | Zinc | |
| | Iron | |
| | Cadmium | |
| | Nickel | |
| | Tin | |
| | Hydrogen | |
| | Copper | |
| | Mercury | |
| | Silver | |
| | Platinum | |
| | Gold | |

## Section 3: REVERSIBILITY OF CHEMICAL REACTIONS

**1** What makes chemical reactions occur, and what affects how rapidly and completely they take place? Several important considerations will be analyzed.

**2** The first question that must be asked is whether a given reaction needs energy or produces energy. Each chemical reaction either uses or gives off a certain quantity of ______________. If energy, usually in the form of heat, is required, the reaction is said to be *endergonic*. If energy is released, the reaction is said to be *exergonic*. The burning of paper or fuel gives off energy (as heat), so it is an ______________ reaction.

energy

exergonic

**3** If an endergonic reaction absorbs energy, that energy must become stored, because we know from the Law of Conservation of Energy that it cannot be destroyed. Endergonic reactions, therefore, result in an increase of *potential/kinetic* energy in the system.

potential

**4** When an endergonic reaction is stated as

$$\mathrm{A + B + energy \longrightarrow C + D} \qquad \text{(III-12)}$$

we will see that C + D, the products, must contain *less/more* energy than the reactants, A + B.

more

**5** If the internal energy of the reactants is $E_1$ and the internal energy of the products is $E_2$, then the change of energy in the system during the reaction is $E_2 - E_1$, or $\Delta E$. The symbol $\Delta$ (delta) refers to the difference in two quantities and is always defined as the final state minus the initial state. $\Delta E =$ ________ $-$ ________.

$E_2 - E_1$

**6** Thus, $\Delta E$ tells us how much ______________ was gained or lost during a reaction.

energy

**7** Returning to the equation given earlier,

$$\mathrm{A + B + energy \longrightarrow C + D} \qquad \text{(III-13)}$$

we see that $E_2$, the energy of the products, is *larger/smaller* than $E_1$. Thus, $\Delta E$ will have a positive value. A positive value for $\Delta E$ tells us that a reaction is *exergonic/endergonic*.

larger

endergonic

**8** On the other hand, a reaction in which

$$\mathrm{A + B \longrightarrow C + D + energy} \qquad \text{(III-14)}$$

means that $E_1$ is *larger/smaller* than $E_2$. The value of $\Delta E$ will thus be *negative*. A negative $\Delta E$ tells us that the reaction is ______________.

larger

exergonic

**9** Reactions may also be classed as *reversible* or *irreversible*. We have already considered one reversible reaction, the partial dissociation of $NH_4OH$:

$$\mathrm{NH_4OH \rightleftharpoons NH_4^+ + OH^-} \qquad \text{(III-15)}$$

Because this reaction goes both ways, it never goes to completion; instead, it establishes a state of equilibrium.

**10** The electrolysis of $CuCl_2$ is an irreversible reaction.

$$\mathrm{Cu^{++} + 2\,Cl^- \longrightarrow Cu^0 + Cl_2\uparrow} \qquad \text{(III-16)}$$

Because both products are removed from the field of activity, the copper as the insoluble metal and the chloride as a gas that escapes, this reaction goes to ______________.

completion

**11** Any reaction that produces a product that cannot participate further will be ______________.

irreversible

chlorine

dissociates

irreversible

**12** This happens if the product is *insoluble,* like copper metal; a *gas,* like ________________; or *nondissociating,* like water. For example, the reaction

$$NaOH + HCl \longrightarrow NaCl + H_2O \qquad \text{(III-17)}$$

goes to completion because water ________________ only slightly.

**13** Because both $OH^-$ and $H^+$ are continuously withdrawn from the field of action, the above reaction is ________________.

**14** Reversible reactions may be made to go to completion by removing one of the products, by increasing the concentration of one of the reactants, or by altering the temperature or pressure of the reaction.

**15** *Le Chatelier's principle* explains the use of temperature and pressure. It states that when a system at equilibrium is subjected to some stress, the equilibrium shifts to relieve the stress.

**16** In the reaction

$$A + B + \text{energy} \rightleftharpoons C + D \qquad \text{(III-18)}$$

endergonic

exergonic

endergonic

remove heat

decreases

principle

the forward reaction (left to right) is ________________, whereas the backward reaction is ________________. Application of heat to this reaction creates stress, which the forward reaction relieves by using up the heat. Application of heat favors the *endergonic/exergonic* pathway. To drive an exergonic reaction, we can *apply heat/remove heat.*

**17** In a reaction involving changes in pressure, applying more pressure favors the reaction that *decreases/increases* the pressure.

**18** In pushing a reversible reaction toward completion by the use of heat or pressure, we are applying Le Chatelier's ________________.

## Section 4: REACTION RATES

**1** The rate of a reaction is the speed at which it progresses. The study of reaction rates is called *kinetics.*

rate

**2** Several factors affect the kinetics, or ____________, of any reaction. Let us look first at a reaction among ions:

$$AgNO_3 + NaCl \longrightarrow NaNO_3 + AgCl \qquad \text{(III-19)}$$

Rewritten in terms of free ions, the equation becomes

$$Ag^+ + NO_3^- + Na^+ + Cl^- \longrightarrow Na^+ + NO_3^- + AgCl\downarrow \qquad \text{(III-20)}$$

irreversible

This reaction involves the chance encounter of an $Ag^+$ and a $Cl^-$ to form the *insoluble* AgCl. Thus, this reaction is *reversible/irreversible.*

$Cl^-$

**3** The rate of a reaction is proportional to the concentrations of the reactants. Thus, if we increase the amount of the reactants $Ag^+$ and ____________, the rate increases. This is so because the more ions there are, the more frequently collisions will occur.

**4** The speed with which small particles such as ions, atoms, or molecules move can be increased by increasing the temperature. Any collisions

increase in speed will also increase the chance of ________ between particles.

**5** When we look at reactions among *molecules,* the situation becomes more complicated. For example, in the reaction

$$H_2 + I_2 \longrightarrow 2\ HI \qquad \text{(III-21)}$$

iodine

it is necessary to separate the 2 atoms of hydrogen and the 2 atoms of ________ from each other before they can react to form HI. This process requires *activation energy,* which can be supplied as heat.

activation

**6** Although the burning of paper is exergonic, the reaction will not start until a lighted match is provided as a source of ________ energy.

activation energy

**7** In the case of an endergonic reaction, some portion of the added heat provides activation energy, and the rest drives the reaction. In general, for each 10° C rise in temperature, the reaction rate doubles. So we can see that, if chemical bonds must be broken before a reaction can take place, both endergonic and exergonic reactions may require ________ ________.

activation

energy

**8** The amount of activation energy required may be lowered by the participation of a *catalyst.* A catalyst is an agent that changes the rate of a reaction without being chemically changed itself. The reaction rate is changed, usually accelerated, because the ________ ________ requirement is lowered.

catalysts

**9** To speed up chemical reactions at temperatures that living things can tolerate, living cells possess special types of ________ called *enzymes.*

concentration

temperature; catalyst

**10** To summarize, the rates of reactions can be increased in three ways: increase the ________ of the reactants, elevate the ________ of the system, or add a(n) ________.

PART **IV**

# Descriptive chemistry of some common compounds

## Section 1: HALIDES

1 The Group VIIA elements in the periodic table are called halogens. They are all *strongly/weakly* electronegative.

strongly

2 Halogens are very reactive and form a variety of compounds. A halide is a compound of a metallic cation with a halogen ion, for example, NaCl, HF, AgI, or KBr.

3 Halides contain 1 or more ions of ________, ________, ________, or ________, bound to a cation.

fluoride; chloride
bromide; iodide

4 Most halides dissociate freely in water to yield their respective ions, such as:

$$NaCl \longrightarrow Na^{+} + Cl^{-} \qquad \text{(IV-1)}$$

A halide compound usually produces two types of ions in water: metallic ________ and halide ________.

cations; anions

## Section 2: OXIDES

1 An *oxide* is a compound containing two elements, one of which is always oxygen. They are named by giving first the name of the major element, then oxide, with a prefix for the number of oxygen atoms (mono-, di-, tri-, tetra-, penta-, and so forth). Examples are

| | |
|---|---|
| CO | carbon *mon*oxide |
| $CO_2$ | carbon *di*oxide |
| BaO | ________ ________ |
| $P_2O_5$ | ________ ________ |

barium oxide
phosphorus pentoxide

The prefix mono- is used only when there are other oxides of the same element.

## Section 3: ACIDS

1 An acid can be defined in several ways. We will use the Brönsted-Lowry concept of acid as a *proton donor*. A proton is the same as a *hydrogen ion* ($H^{+}$); it is the nucleus of a ________ atom.

hydrogen

hydrogen ion

**2** In aqueous solution the free proton, or __________ ______, undergoes a further reaction with water:

$$H^+ + H_2O \longrightarrow H_3O^+ \quad \text{(IV-2)}$$

The product is called *hydronium* ion. Thus, acids in aqueous solution are characterized by the presence of $H_3O^+$.

**3** Oxides of nonmetals react with water to form acids; for example:

$$SO_3 + H_2O \longrightarrow H_2SO_4 \quad \text{(IV-3)}$$

sulfur trioxide

This equation reads $SO_3$, or __________ __________, reacts with water to yield sulfuric acid.

**4** We know the product $H_2SO_4$ is an acid because it reacts further with water, as follows:

$$H_2SO_4 + 2\ H_2O \longrightarrow 2\ H_3O^+ + SO_4^{=} \quad \text{(IV-4)}$$

hydronium

yielding __________ ions, $H_3O^+$.

**5** The equation for the formation of an acid by carbon dioxide is:

$CO_2 + H_2O \longrightarrow H_2CO_3$

______ + ______ $\longrightarrow$ ______ (IV-5)

carbon dioxide — carbonic acid

**6** The equation for the reaction between carbonic acid and water is:

$H_2CO_3$; $H_3O^+$

______ $+ H_2O \longrightarrow$ ______ $+ HCO_3^-$ (IV-6)

nonmetallic

**7** Sulfuric acid, carbonic acid, and other acids are formed in reactions between __________ oxides and water. Such oxides in their "dry state" are called *acid anhydrides,* which means without water. Therefore, a compound such as $NO_2$ is not only a nonmetallic oxide, but is chemically an __________ __________.

acid anhydride

**8** Other common acids are formed when the *hydrogen halides* HCl, HF, HBr, and HI dissolve in water. For example, hydrogen chloride reacts with water:

$$HCl + H_2O \longrightarrow H_3O^+ + Cl^- \quad \text{(IV-7)}$$

The acid solution that results is called hydrochloric acid.

**9** The concepts of complete and partial dissociation were discussed in Part III and can be reviewed if necessary.

**10** Acids may be classed as *strong acids* or *weak acids,* on the basis of the extent to which they __________ in water to yield hydronium ions.

dissociate

**11** Sulfuric acid ($H_2SO_4$) undergoes essentially complete dissociation:

$$H_2SO_4 + 2\ H_2O \longrightarrow 2\ H_3O^+ + SO_4^{=} \quad \text{(IV-8)}$$

completion

The single arrow denotes a reaction that runs to __________. Because virtually every molecule of $H_2SO_4$ dissociates, it is a strong acid.

**12** On the other hand, nitrous acid ($HNO_2$) dissociates weakly:

$$HNO_2 + H_2O \rightleftharpoons H_3O^+ + NO_2^- \quad \text{(IV-9)}$$

reversible

weak

The double arrow indicates a __________ reaction; thus, nitrous acid is a *strong/weak* acid.

**Table 5.** Common inorganic acids

| *Empirical formula* | *Acid* |
|---|---|
| HF | Hydrofluoric |
| HCl | Hydrochloric |
| HBr | Hydrobromic |
| HI | Hydriodic |
| $H_2S$ | Hydrosulfuric |
| $HNO_2$ | Nitrous |
| $HNO_3$ | Nitric |
| $H_2SO_3$ | Sulfurous |
| $H_2SO_4$ | Sulfuric |
| $H_3PO_4$ | Phosphoric |
| $H_2CO_3$ | Carbonic |

**13** As a general rule, inorganic acids are named as follows: binary acids, those composed of hydrogen and one other element, are named using the prefix *hydro-* and suffix *-ic;* ternary acids, those composed of hydrogen, oxygen, and another element, carry the stem name of the characteristic third element plus the suffix *-ous* or *-ic* to indicate respectively the lower and higher amount of oxygen.

binary
HF

**14** Hydrofluoric acid is a ______________ acid with the formula ________.

ternary
nitric acid

**15** $HNO_2$ and $HNO_3$ are ______________ acids with the names nitrous acid and ______________ ______________ respectively.

**16** Common important inorganic acids are presented in Table 5.

## Section 4: BASES

**1** *Bases* are compounds that dissociate in water to furnish *hydroxyl* ions ($OH^-$). They may also be defined under the Brönsted-Lowry concept as compounds that act as *proton acceptors.*

**2** The first type of base is prepared by reacting a *metallic* oxide with water. For example:

$$K_2O + H_2O \longrightarrow 2\ KOH \qquad \text{(IV-10)}$$

base

The product, potassium hydroxide, is a ______________. Metallic oxides are called *basic anhydrides.*

**3** KOH dissociates in water as follows:

$$KOH \longrightarrow K^+ + OH^- \qquad \text{(IV-11)}$$

hydroxyl

releasing the characteristic ______________ ion.

**4** The hydroxyl ion is a *proton acceptor* if $H^+$ is present:

$$OH^- + H^+ \longrightarrow H_2O \qquad \text{(IV-12)}$$

$OH^-$

**5** Other compounds may act as bases without releasing the ________ ion. For example, ammonia ($NH_3$) reacts as follows:

$$NH_3 + H^+ \longrightarrow NH_4^+ \qquad \text{(IV-13)}$$

acceptor — and behaves as a proton ________________, or Brönsted-Lowry base.

**6** Like acids, bases that release the hydroxyl ion may be characterized as strong or weak, depending on the degree of ________________ in water. — dissociation

**7** The dissociation of a strong base, such as NaOH, is complete:

$NaOH \longrightarrow Na^+ + OH^-$

$$\_\_\_\_\_\_ \longrightarrow \_\_\_\_\_\_ + \_\_\_\_\_\_ \qquad \text{(IV-14)}$$

whereas a weak base has partial dissociation, as does $NH_4OH$:

$NH_4OH$ ($\rightleftharpoons$) $NH_4^+ + OH^-$

$$\_\_\_\_\_\_ (\_\_\_) \_\_\_\_\_\_ + \_\_\_\_\_\_ \qquad \text{(IV-15)}$$

basic anhydrides

**8** Naming the bases formed from metallic oxides, or ________________ ________________, simply involves adding the term hydroxide to the name of the cation; thus, NaOH is sodium hydroxide.

**9** The names for the following are:

| | | |
|---|---|---|
| barium hydroxide | $Ba(OH)_2$ | ________________ ________________ |
| ferrous hydroxide | $Fe(OH)_2$ | ________________ ________________ |
| ferric hydroxide | $Fe(OH)_3$ | ________________ ________________ |

## Section 5: SALTS

**1** An acid reacts with a base to form two products. One is a *salt,* the other is water. Thus:

$$HCl + NaOH \longrightarrow NaCl + H_2O \qquad \text{(IV-16)}$$

A salt, then, is a compound of a metal and a nonmetal, usually formed in the reaction between an acid and a base.

**2** An acid-base reaction is called *neutralization,* because it removes both hydroxyl ions and ________________ ________________ from solution. Because water is essentially *undissociated,* neutralization reactions are *reversible/irreversible.* — hydrogen ions; irreversible

**3** There are three methods for preparing salts in the laboratory. In one, we react an acid containing the appropriate anion with a base containing the appropriate ________________. Thus, to make calcium nitrate, we use calcium hydroxide and nitric acid: — cation

$$Ca(OH)_2 + 2\ HNO_3 \longrightarrow Ca(NO_3)_2 + 2\ H_2O \qquad \text{(IV-17)}$$

**4** In the second, we react an acid with a basic anhydride, or a ________________ oxide, possessing the desired cation. Thus: — metallic

$$CaO + 2\ HNO_3 \longrightarrow Ca(NO_3)_2 + H_2O \qquad \text{(IV-18)}$$

This is feasible because CaO in water forms calcium hydroxide:

$H_2O$; $Ca(OH)_2$

$$CaO + \_\_\_\_\_\_ \longrightarrow \_\_\_\_\_\_ \qquad \text{(IV-19)}$$

**5** In the third method, salts may be formed in reactions between certain metals and acids. In Table 4, hydrogen is included among the ________________ elements. Note its relative position in regard to replacement activity. Hydrogen can be replaced from an acid by — metallic

| | |
|---|---|
| iron | any metal *above it* in the series. Thus, hydrogen can be replaced by *iron/silver*. |

**6** Therefore, in the reaction between aluminum and sulfuric acid

$$2\ Al + 3\ H_2SO_4 \longrightarrow Al_2(SO_4)_3 + 3\ H_2 \quad \text{(IV-20)}$$

| | |
|---|---|
| hydrogen<br>salt | aluminum replaces ______________ from the acid and forms a ______________. |

**7** Salts may react with strong bases; for example:

$$FeCl_3 + 3\ NaOH \longrightarrow Fe(OH)_3 + 3\ NaCl \quad \text{(IV-21)}$$

| | |
|---|---|
| dissociated | This and related reactions progress to completion only because the new base formed, in this case $Fe(OH)_3$, is weakly ______________. |

**8** Salts may react with other salts; for example:

$$MgSO_4 + BaCl_2 \longrightarrow MgCl_2 + BaSO_4 \quad \text{(IV-22)}$$

| | |
|---|---|
| insoluble | These reactions progress to completion only if one of the products is ______________ or a precipitate, such as $BaSO_4$. |

**9** Salts may react with acids; for example:

$$2\ NaNO_3 + H_2SO_4 \longrightarrow Na_2SO_4 + 2\ HNO_3\uparrow \quad \text{(IV-23)}$$

| | |
|---|---|
| salt | The products are an acid and a(n) ______________. This reaction can go to completion, because nitric acid can be distilled off as it is formed. |

**10** Salts are named by combining the name of the cation with the name of the anion (see Table 3, p. 12). The names for the following are:

| | | |
|---|---|---|
| sodium iodide | NaI | ______________ ______________ |
| ammonium sulfide | $(NH_4)_2S$ | ______________ ______________ |
| magnesium sulfate | $MgSO_4$ | ______________ ______________ |
| barium nitrite | $Ba(NO_2)_2$ | ______________ ______________ |

## Section 6: ACID AND BASIC SALTS

**1** *Acid salts* are formed when an acid is only partially neutralized. Thus, when sulfuric acid and potassium hydroxide react, either

$$H_2SO_4 + KOH \longrightarrow KHSO_4 + H_2O \quad \text{(IV-24)}$$

or

$$H_2SO_4 + 2\ KOH \longrightarrow K_2SO_4 + 2\ H_2O \quad \text{(IV-25)}$$

| | |
|---|---|
| hydrogen | will occur, depending on the proportions of the reactants. $KHSO_4$ is called an acid salt because it retains 1 ______________ atom per molecule. |

**2** The formulas for the three salts that may be formed in the reaction of $H_3PO_4$ + NaOH are:

| | | |
|---|---|---|
| $NaH_2PO_4$ | Monosodium phosphate | ______________ |
| $Na_2HPO_4$ | Disodium phosphate | ______________ |
| $Na_3PO_4$ | Sodium phosphate | ______________ |

**3** *Basic salts* are formed when a base is only partially neutralized.

When lead hydroxide and nitric acid react, either

$$Pb(OH)_2 + HNO_3 \longrightarrow Pb(OH)NO_3 + H_2O \qquad \text{(IV-26)}$$

or

$$Pb(OH)_2 + 2\ HNO_3 \longrightarrow Pb(NO_3)_2 + 2\ H_2O \qquad \text{(IV-27)}$$

may occur. $Pb(OH)NO_3$, basic lead nitrate, is a basic salt because it retains a(n) ________________ ion. — hydroxyl

**4** The formulas for the three salts that may be formed in the reaction between $[Bi(OH)_3] + HCl$ are:

________________ — $Bi(OH)_2Cl$

________________ — $Bi(OH)Cl_2$

________________ — $BiCl_3$

## Section 7: AMPHOTERIC COMPOUNDS

**1** An *amphoteric* compound can form salts with both bases and acids; that is, when combined with an acid it acts like a(n) ____________, and when combined with a base it acts like a(n) ____________. — base; acid

**2** Within the Brönsted-Lowry concept of acids and bases, an acid is a proton ____________, and a base is a ____________ ____________. — donor; proton acceptor

**3** Let us look at the behavior of $Zn(OH)_2$, an amphoteric compound. With an acid, it reacts as follows:

$$Zn(OH)_2 + 2\ HCl \longrightarrow ZnCl_2 + 2\ H_2O \qquad \text{(IV-28)}$$

Here the hydroxyl ion ($OH^-$) functions as a proton ____________, and $Zn(OH)_2$ is by definition acting as a(n) ____________. — acceptor; base

**4** With a strong base, such as NaOH, $Zn(OH)_2$ reacts as follows:

$$Zn(OH)_2 + 2\ NaOH \longrightarrow Na_2ZnO_2 + 2\ H_2O \qquad \text{(IV-29)}$$

Here the hydroxyl ion of the strong base, NaOH, has accepted $H^+$ from $Zn(OH)_2$ to form water. Thus $Zn(OH)_2$ functions as an acid, because it plays the part of a ____________ ____________. — proton donor

**5** Water itself can also function as an amphoteric compound; that is, it can be either a proton donor or ____________. — acceptor

**6** We are already familiar with the reaction where protons, released by some acid, combine with water:

$$H^+ + H_2O \longrightarrow H_3O^+ \qquad \text{(IV-30)}$$

This yields the ________________ ion. In this reaction, water acts as a(n) *acid/base*. — hydronium; base

**7** We can expand our understanding of the chemical behavior of water by stating that water itself undergoes very slight dissociation (about 1 in every 10 million molecules) as follows:

$$H_2O \rightleftharpoons H^+ + OH^- \qquad \text{(IV-31)}$$

Thus, when sodium acetate ($NaC_2H_3O_2$) is dissolved in water, we have

$$Na^+ + C_2H_3O_2^- + H_2O \rightleftharpoons HC_2H_3O_2 + OH^- + Na^+ \qquad \text{(IV-32)}$$

donates · acid · base

Here water ______________ protons to the acetate ion ($C_2H_3O_2$) and can be said to function as a(n) *acid/base;* the acetate ion functions as a(n) __________.

## Section 8: HYDROLYSIS

1 Many salts, in addition to dissociating in water, also undergo reactions with water. Any chemical reaction in which water is a reactant is called *hydrolysis*. The occurrence of hydrolysis explains why aqueous solutions of many salts are not neutral. Consider four classes of salts.

2 Some salts are formed from *strong* acids and *strong* bases; for example, NaCl is formed from HCl and NaOH:

$$HCl + NaOH \longrightarrow NaCl + H_2O \quad \text{(IV-33)}$$

hydrolysis

Salts of strong acids and strong bases are neutral when dissolved in water, because they do not react with water, or undergo __________. Thus:

$$NaCl + H_2O \longrightarrow Na^+ + Cl^- + H_2O \quad \text{(IV-34)}$$

gives no excess of free protons or free hydroxyl ions.

3 Some salts, such as $NH_4Cl$, are formed from *strong* acids and *weak* bases:

$$HCl + NH_4OH \longrightarrow NH_4Cl + H_2O \quad \text{(IV-35)}$$

dissociated

$NH_4OH$ is a weak base because it is only slightly ____________ in water.

4 Salts of strong acids and weak bases dissolve in water to give acidic solutions. In the reaction between $NH_4Cl$ and water, the salt dissociates as follows:

$$NH_4Cl \longrightarrow NH_4^+ + Cl^- \quad \text{(IV-36)}$$

Water also normally undergoes very slight dissociation:

$H^+ + OH^-$

$$H_2O \rightleftharpoons \_\_\_\_ + \_\_\_\_ \quad \text{(IV-37)}$$

Combined, these two simultaneous dissociations are:

$NH_4^+ + Cl^- +$ $H^+ + OH^-$

$$NH_4Cl + H_2O \rightleftharpoons \_\_\_\_ + \_\_\_\_ + \_\_\_\_ + \_\_\_\_ \quad \text{(IV-38)}$$

5 Remembering that

$$NH_4^+ + OH^- \rightleftharpoons NH_4OH \text{ (a weak base)} \quad \text{(IV-39)}$$

protons; hydroxyl ions · $NH_4OH$

we see that although dissociation of water produces equal numbers of ____________ and ____________ __________, the latter are largely removed in the formation of the weak base __________.

protons

6 An aqueous solution of the salt of a strong acid and weak base (such as $NH_4Cl$) is acidic because of an excess of ____________.

protons

7 The hydrolysis reaction of $NH_4Cl$ with water will result in an excess of *protons/hydroxyl ions*.

**8** A third class of salts are those of *weak* acids and *strong* bases. For example, sodium acetate, $NaC_2H_3O_2$ is formed as follows:

$$HC_2H_3O_2 + NaOH \longrightarrow NaC_2H_3O_2 + H_2O \qquad \text{(IV-40)}$$

Such salts form *basic* solutions.

**9** The combined reaction for the dissociation of the salt and water is:

$Na^+ + C_2H_3O_2^- + H^+ + OH^-$

$$NaC_2H_3O_2 + H_2O \rightleftharpoons \_\_\_\_ + \_\_\_\_ + \_\_\_\_ + \_\_\_\_ \qquad \text{(IV-41)}$$

Because

$HC_2H_3O_2$

$$C_2H_3O_2^- + H^+ \rightleftharpoons \_\_\_\_ \text{ (a weak acid)} \qquad \text{(IV-42)}$$

hydroxyl

basic

more protons are removed from solution than ________ ions. The net result is a(n) *acid/basic* solution.

**10** The fourth class of salts are those of *weak* acids and *weak* bases, for example, ammonium acetate ($NH_4C_2H_3O_2$). Such salts form neutral solutions. When $NH_4C_2H_3O_2$ is dissolved in water, we have:

$NH_4^+ + C_2H_3O_2^- + H^+ + OH^-$

$$NH_4C_2H_3O_2 + H_2O \rightleftharpoons \_\_\_\_ + \_\_\_\_ + \_\_\_\_ + \_\_\_\_ \qquad \text{(IV-43)}$$

hydroxyl

protons

Here $NH_4^+$ tends to remove ________ ions, and $C_2H_3O_2^-$ tends to remove ________. Because these two effects are balanced, the solution is neutral.

neutral; acidic

basic

neutral

**11** To summarize, in solutions the salts of strong acids and strong bases are ________, salts of strong acids and weak bases are ________, those of weak acids and strong bases are ________, and of weak acids and weak bases are ________.

hydrolysis

**12** Any salt that reacts with water is said to undergo ________.

protons

hydroxyl ions

**13** When weak acids are formed in hydrolysis, ________ are withdrawn from solution, whereas when weak bases are formed, ________ ________ are withdrawn.

# PART V

# Stoichiometry

## Section 1: MOLECULAR WEIGHTS

**1** *Stoichiometry* is quantitative chemistry. It involves the determination of the exact proportions of the combining elements in any chemical reaction. When we want to know how much of a given compound will be used in a reaction, we must have an understanding
stoichiometry — of ________________.

**2** Each element has an *atomic weight,* which is the average weight of 1 atom. Thus the atomic weight of boron, rounded off to the nearest whole number, is 11. The atomic weights of oxygen and chromium
16; 52 — are _______ and _______ respectively.

**3** The molecular weight of a compound is the weight of an average molecule. It is computed by adding up the atomic weights of the constituent atoms as indicated by the formula.

**4** The molecular weight of water, $H_2O$, is calculated simply by adding the atomic weights of the atoms:

| | | |
|---|---|---|
| H = | 1 × 2 (number of atoms) = | 2 |
| O = | 16 × 1 = | 16 |
| Total | | 18 |

molecular — Thus, the ________________ weight of water is 18.

**5** The molecular weights of the following are:

| | | |
|---|---|---|
| 36.5 | HCl | _______ |
| 44 | $CO_2$ | _______ |
| 99.4 | $Zn(OH)_2$ | _______ |
| 342 | $Al_2(SO_4)_3$ | _______ |

**6** Chemists frequently use a convenient unit called the *mole*. One mole of a compound is 1 *gram-molecular weight* (GMW). For example, 1 mole of water equals 18 g of water (the molecular weight in grams).
40 — One mole of NaOH is _______ g.

1.5 — **7** 60 g of NaOH equals _______ moles.

**8** A mole of any compound contains $6.02 \times 10^{23}$ molecules of that compound. This is called *Avogadro's number*. Thus, 1 mole of $NH_3$,
17; 160 — weighing _______ grams, and 1 mole of $CuSO_4$, weighing _______
$6.02 \times 10^{23}$ — grams, each contain ________________ molecules.

## Section 2: PERCENT COMPOSITION

**1** It is often useful to know what percent of a molecule is represented by one of its elements. For example, what percent of $CO_2$ is carbon?

**2** Percent composition is readily calculated by the arithmetical proportion

$$\frac{\text{part}}{\text{whole}} \times 100 = \text{percent}$$

Thus, to solve for percent carbon in $CO_2$ we substitute the atomic weight of carbon as the *part* and the molecular weight of $CO_2$ as the

whole ______________:

$$\frac{12}{44} \times 100 = 27\%$$

**3** To solve for percent oxygen in $CO_2$ we use the atomic weight of oxygen as the part and multiply by 2 because there are 2 oxygen atoms:

73% $$\frac{2 \times 16}{44} \times 100 = \_\_\_\_\_\_\_\_\%$$

**4** We can check our calculations by confirming that the *parts* add up to

100 the *whole;* that is, 27% carbon plus 73% oxygen equals ________%.

**5** The percentages of the following are:

2.7% H in HCl ______________

27.1% Na in $NaNO_3$ ______________

43.2% O in $Ca(OH)_2$ ______________

14.3% H in $NH_4OH$ ______________

## Section 3: WEIGHT-WEIGHT PROBLEMS

**1** This type of calculation deals with finding out what weight of one chemical it takes to react with a given weight of another chemical, or to yield a given weight of product. To start with, we need a balanced equation.

**2** Consider the reaction between sodium bicarbonate and sulfuric acid. First we write the balanced equation:

$$2\ NaHCO_3 + H_2SO_4 \longrightarrow Na_2SO_4 + 2\ H_2O + 2\ CO_2$$

If we wish to compute the weight of $H_2SO_4$, which reacts completely with 100 g of $NaHCO_3$, we first must calculate the molecular weights of the two reactants:

84 $NaHCO_3$ ________

98 $H_2SO_4$ ________

**3** Substitute these values as follows and solve for X:

$$\frac{\text{g } NaHCO_3}{\text{coefficient} \times \text{mol. wt. } NaHCO_3} = \frac{\text{g } H_2SO_4}{\text{coefficient} \times \text{mol. wt. } H_2SO_4}$$

$$\frac{100}{2 \times 84} = \frac{X}{98}$$

$$X = 58 \text{ g}$$

weight
weight

4 The answer is expressed in *grams* because this is a __________-__________ problem.

5 To solve for the weight of $Na_2SO_4$ (X) produced in the above reaction, *first* calculate its molecular weight. Then set up one of two proportions:

$$\frac{100}{168} = \frac{X}{(\text{mol. wt. } Na_2SO_4)}$$

or

142

$$\frac{58}{98} = \frac{X}{\phantom{142}}$$

84

6 In either of the above proportions, X equals ________ g.

7 In calculating the weight of $ZnCl_2$ formed in the reaction between 25 g of Zn and sufficient HCl, the equation is

$$Zn + 2\ HCl \longrightarrow ZnCl_2 + H_2$$

and the proportion is

$\frac{25}{65} = \frac{X}{135}$

$$\frac{\phantom{25}}{\phantom{65}} = \frac{X}{\phantom{135}}$$

51.9

8 The weight of $ZnCl_2$ formed in the above reaction is ________ g.

9 The equation used in steel manufacture is

$$3\ CO + Fe_2O_3 \longrightarrow 3\ CO_2 + 2\ Fe$$

39.4

The weight of CO needed to reduce 75 tons of $Fe_2O_3$ is ________ tons.

*Calculation:*

$\frac{X}{3(28)} = \frac{75}{160}$

## Section 4: WEIGHT-VOLUME PROBLEMS

mole

1 When a gas is either a reactant or product in a reaction, we usually solve for the *volume* of the gas rather than its weight. To do this we use *Avogadro's law*. This law states that, at the same temperature and pressure, *equal volumes* of all gases contain the *same number* of molecules. We already know that 1 __________ of any pure element or compound contains Avogadro's number ($6.02 \times 10^{23}$) of atoms or molecules.

temperature; pressure

2 Avogadro's law tells us that temperature and pressure conditions are important when doing calculations with gases. Therefore, in doing weight-volume problems, we will use *standard conditions* of __________ and __________. This means the temperature is 0° C, and the pressure is 760 mm Hg (sea level atmospheric pressure).

0; 760

3 One mole of any gas at standard conditions, ________ ° C and ________ mm Hg, occupies a volume of exactly 22.4 liters. This is called the *gram-molecular volume*. Substitute this value for the molecular weight of the gas in solving weight-volume problems.

gram-molecular **4** When we use the ____________-________________ volume, we are referring to $6.02 \times 10^{23}$ molecules of the gas, just as we used the mole to refer to this value.

standard **5** All weights in the weight-volume problems must be expressed in *grams;* all volumes must be expressed in *liters* at ____________ conditions.

**6** To calculate the volume of $CO_2$ formed from 100 g of $NaHCO_3$ in the reaction

$$2\ NaHCO_3 + H_2SO_4 \longrightarrow NaSO_4 + 2\ H_2O + 2\ CO_2$$

we set up the proportion

$$\frac{100}{168} = \frac{X}{2\,(22.4)}$$

Note that the gram-molecular volume replaces the molecular weight of $CO_2$.

**7** Solving for X in the above proportion, we obtain 26.6 liters of $CO_2$.

34.7 **8** The volume of $CO_2$ formed when 76 g of $H_2SO_4$ reacts with a sufficient amount of $NaHCO_3$ is ______ liters of $CO_2$.

*Calculation:*

$$\frac{76}{98} = \frac{X}{2(22.4)}$$

**9** In the reaction

$$Zn + 2\ HCl \longrightarrow ZnCl_2 + H_2\uparrow$$

17.2 the volume of hydrogen gas released when 50 g of Zn reacts with sufficient HCl is ______ liters $H_2$.

*Calculation:*

$$\frac{50}{65} = \frac{X}{22.4}$$

## Section 5: VOLUME-VOLUME PROBLEMS

**1** Volume-volume problems are simple. We use *Gay-Lussac's law* in solving them, which states that when gases react, their relative combining volumes and the volumes of the products, if gases, can be expressed in *small whole numbers*. These correspond to the *coefficients* in the equation.

**2** For example, in the Haber process for making ammonia:

$$N_2 + 3\ H_2 \longrightarrow 2\ NH_3$$

Gay-Lussac's All participants in this reaction are gases, so ____________-____________ law can be applied.

**3** Thus, if we start with 10 liters of $N_2$ and wish to calculate the volume of $NH_3$ produced, we set up a proportion using the coefficients in the equation:

$$\frac{10\ (\text{liters of } N_2)}{1\ (\text{coefficient of } N_2)} = \frac{X\ (\text{liters of } NH_3)}{2\ (\text{coefficient of } NH_3)}$$

$$X = 20\ \text{liters}$$

90

**4** The volume of hydrogen required to yield 60 liters of $NH_3$ is ______ liters of $H_2$.

*Calculation:*

$\frac{X}{3} = \frac{60}{2}$

**5** In the reaction

$$2\,NO + O_2 \longrightarrow 2\,NO_2$$

the volume of $NO_2$ produced from 30 liters of NO (with sufficient

30

$O_2$) is ______ $NO_2$.

*Calculation:*

$\frac{30}{2} = \frac{X}{2}$

15

**6** In the above problem, sufficient oxygen is ______ liters $O_2$.

*Calculation:*

$\frac{X}{1} = \frac{30}{2}$

PART **VI**

# Gas laws

## Section 1: THE GASEOUS STATE

1 We saw in the first part of this program that matter can exist in three states: __________, ______________, and ________. We will be concerned here with the special properties of gases and certain natural laws they obey.

solid; liquid; gas

2 Remember that gases have neither a fixed shape nor a fixed ________.

volume

3 The volume of a gas may be affected by conditions of temperature and ______________.

pressure

## Section 2: BOYLE'S LAW

1 Boyle's law describes the relationship between the *pressure* on a gas and its volume. When we increase the pressure on a gas, such as the vapor in an automobile engine cylinder, its volume *decreases*. Conversely, when we decrease pressure, the volume of the gas increases. This is called an *inverse* relationship. Boyle's law states that the volume of a gas varies inversely with the pressure if the temperature remains constant.

2 Boyle expressed the relationship mathematically as

$$\frac{V'}{V} = \frac{P}{P'}$$

where V and P are the starting volume and pressure respectively and $V'$ and $P'$ are the ending ______________ and ______________.

volume; pressure

3 Gas volumes in chemistry and physics are expressed in liters (l) and milliliters (ml) or cubic centimeters (cc) and commercially in cubic feet. Pressures are expressed in millimeters of mercury (mm Hg). Sometimes we use *atmospheres* as pressure measurements. One atmosphere is equal to 760 mm Hg, so 0.5 atmosphere equals ______ mm Hg.

380

4 Suppose under a pressure of 750 mm Hg a gas has a volume of 425 ml. If the pressure is dropped to 375 mm Hg, what volume will the gas occupy? From our preliminary discussion we can predict that the second volume will be *larger/smaller* than the first.

larger

**5** Substituting these values in the Boyle's law equation and solving for V′, we have

$$\frac{V'}{425\ ml} = \frac{750}{375}$$

850 $V' =$ ________ ml

**6** A gas in a 20-liter vessel has a pressure of 142 mm Hg. If the gas is transferred to another vessel and its new pressure is 360 mm, and there is no change in temperature, then its volume in the new vessel is

8 ________ liters.

*Calculation:*

$$\frac{V'}{20} = \frac{142}{360}$$

## Section 3: CHARLES'S LAW

**1** Charles's law describes the relationship between the temperature of a gas and its volume. In warming a balloon, we observe that it gets larger; that is, the volume of the air inside increases. This is a *direct* relationship. Charles's law states that the volume of a gas is *directly proportional* to its absolute temperature, if the pressure remains constant.

**2** Charles expressed this mathematically as:

$$\frac{V}{V'} = \frac{T}{T'}$$

volume where V and T are the starting ________________ and

temperature ________________, V′ and T′ the final values.

**3** When dealing with gases we must use the absolute temperature scale, expressed in kelvins (K) (degree sign not used). Absolute temperature is obtained by adding 273 to the Celsius reading. Thus 10° C = 283 K; −20° C = ________ K.

253 K

**4** If 100 ml of $O_2$ at 283 K were cooled to 223 K, its new volume at the same pressure would become *larger/smaller*.

smaller

**5** Substituting the above values in Charles's equation and solving for V′, we have:

$$\frac{V}{V'} = \frac{T}{T'}$$

$$\frac{100}{V'} = \frac{283}{223}$$

$$V' = 79\ ml$$

**6** If 35 ml of a gas were subjected to a temperature change from 65° C to 95° C, then its new volume at the same pressure is

38 ________ ml.

*Calculation:*

$$\frac{35}{V'} = \frac{338}{368}$$

## Section 4: THE COMBINED GAS LAW

Boyle's
Charles's

1 In most systems involving gases, *both* pressure and temperature change simultaneously. The results then relate to both ________ and ________ laws. For example, when you open a cold bottle of a carbonated beverage in a warm room, the effervescense results from both the release of (drop in) pressure and the increase in temperature.

volume (V)

2 The combined gas law equation incorporates both Boyle's and Charles's equations and states the relationships among three measured properties of a gas: temperature (T), pressure (P), and ________.

combined

3 The ________ gas law equation is written:

$$\frac{PV}{T} = \frac{P'V'}{T'}$$

461

4 If a gas at 313 K and 550 mm Hg occupies a volume of 350 ml, what pressure is required to bring its volume to 400 ml at 300 K? Substituting these values in the combined gas law equation and solving for P′ we have ______ mm Hg.

*Calculation:*

$$\frac{550 \times 350}{313} = \frac{P' \times 400}{300}$$

147

$$\frac{1 \times 200}{293} = \frac{10 \times 10}{T'}$$

5 If a gas occupies a volume of 200 l at 20° C and 1 atmosphere, the temperature required to bring it to 10 l and 10 atmospheres is ______ K.

*Calculation:*

## Section 5: DALTON'S LAW

1 Frequently gases are found in mixtures, such as in air. Simplified, *Dalton's law of partial pressures* states that the total pressure of a mixture of nonreacting gases is the sum of the *partial* pressures of the constituent gases. By partial pressure we mean the pressure the gas would exert if it existed alone in the same volume as occupied by the mixture.

2 Dalton's law is stated mathematically as

$$P = p_1 + p_2 + p_3 \ldots$$

partial
pressures

P is the total pressure, and $p_1$, $p_2$, $p_3$, and so forth, are the ________ ________ of the constituent gases.

total
pressure

3 A gas in a mixture contributes to the total pressure in proportion to its percentage of the total volume. Thus, a gas that is 50% of a mixture will contribute a partial pressure equal to 50% of the ________ ________.

4 In a mixture containing 40% $CO_2$ and 60% $O_2$ and having a total

200 pressure of 500 mm Hg, $CO_2$ has a partial pressure ($p_1$) of ______
300 mm Hg, and $O_2$ has a partial pressure ($p_2$) of ______ mm Hg.

**5** Air is a mixture of 78% $N_2$, 20% $O_2$, 0.93% argon, 0.035% $CO_2$, and traces of other gases. Thus the partial pressures of $O_2$ ($p_1$) and $N_2$ ($p_2$) at 1 atmosphere pressure are:

152 $p_1$ = ______ mm Hg
593 $p_2$ = ______ mm Hg

PART **VII**

# Solutions

## Section 1: GENERAL CHARACTERISTICS OF SOLUTIONS

**1** A *solution* is any homogeneous mixture where the constituent molecules, atoms, or ions are completely intermingled. Although it is correct to speak of gases dissolved in gases, solids dissolved in solids, and other combinations, in most instances we encounter solutions of the three states dissolved in liquids.

solid
liquid

**2** When we dissolve sugar in water, we are dissolving a ________ in a ________. The constituent present in the larger proportion, the liquid, is called the *solvent;* the lesser constituent is the *solute*.

solute
solvent

**3** Thus, when we dissolve a fat in alcohol, the fat is the ________, and the alcohol is the ________.

soluble

**4** We can dissolve a great deal of NaCl in a small amount of water. Thus we say that NaCl is *highly soluble* in water. If we add $CaCO_3$ to water, only a small amount will dissolve, because $CaCO_3$ is only slightly ________ in water. If we add $Fe(OH)_3$ to water, we do not detect any dissolving action; $Fe(OH)_3$ is practically *insoluble* in water.

dissolve

**5** When a small amount of sugar is dissolved in water, the solution is *unsaturated,* because more sugar will readily ________ if added. As more solute is added, a point is reached at which no more dissolves and excess solute will remain as crystals on the bottom. This is called a *saturated* solution.

equilibrium

**6** As long as undissolved solute is present in a saturated solution, an equilibrium will exist. As 1 molecule of sugar leaves the solution, another dissolves and enters the solution. In a saturated solution, dissolved and undissolved sugar molecules are in ________.

crystallization

**7** Molecules leaving the saturated solution will join the undissolved molecules to form crystals. This process is called *crystallization*. We can say, then, that in a saturated solution at equilibrium the rate of solution equals the rate of ________.

increases

**8** In most cases, heating a saturated solution will cause some excess solute to dissolve. Solubility increases as the temperature *increases/decreases*.

crystallize

**9** If crystals of solute are present when a saturated solution is *cooled,* excess solute will ________ out of solution. If no undissolved solute is present, the excess *may* remain in solution. Such a solution is said to be *supersaturated*.

**10** If a *crystallization nucleus* is provided in a supersaturated solution by adding a crystal of solute, dramatic crystallization will occur, and the solution will revert to the ________________ state. Seeding rain clouds with ice crystals to provoke rainfall is a practical example of the use of the ________________ nucleus.

saturated

crystallization

**11** Although the amount of a *solid* that will dissolve in a liquid increases with rising temperature, the reverse is usually true for a gas dissolved in liquid. As a liquid is cooled, more gas will dissolve. Therefore, lake water in summer contains *less/more* dissolved $O_2$ than in winter.

less

**12** The solubility of a gas in a liquid increases with increasing pressure. If we wish to carbonate beer to give it a good head, we should add $CO_2$ under *low/high* pressure.

high

**13** The solubility of liquids in liquids is referred to as *miscibility*. Liquids that will not mix, such as oil and water, are said to be *immiscible*. Liquids such as alcohol and water, which form true solutions in all proportions, are said to be fully ________________.

miscible

## Section 2: PERCENT CONCENTRATION

**1** Because solutions are mixtures, the components will not have definite proportions, as compounds do. *Concentration* tells us the amount of solute relative to the volume of solution.

**2** The most familiar way to express concentration is percentage by *weight per unit volume (w/v)*. The weight of solute in grams is the *part;* the volume of solution in milliliters is the *whole*. This can be expressed mathematically as

$$\% = \frac{\text{g solute}}{\text{ml solution}} \times 100$$

Thus, if we wish to find the concentration of a solution of 2 g of $ZnCl_2$ in 40 ml of water, we substitute and solve as follows:

$$\% = \frac{2}{40} \times 100$$

$$\% = \_\_\_\_\_\_\_\_$$

5

**3** A solution containing 47 g of NaCl in 350 ml (of solution) contains ________ % NaCl.

13.4

*Calculation:*

$$\% = \frac{47}{350} \times 100$$

**4** To make a 6% solution of $NaNO_3$ with 18 g of a compound, we need enough water to make _______ ml of solution.

300

*Calculation:*

$$6\% = \frac{18}{X} \times 100$$

**5** For solutions of liquids in liquids, we use *volume* of solute as *part,* remembering that the *lesser* component is the solute and volume of solution is the *whole*. The equation is

$$\% = \frac{\text{volume solute}}{\text{volume solution}} \times 100$$

25 A solution containing 50 ml alcohol and 150 ml water is ______ % alcohol.

*Calculation:*

$\% = \frac{50}{200} \times 100$

**6** The volume of glycerol contained in 425 ml of a 15% v/v solution of
64 glycerol in water is ______ ml.

*Calculation:*

$15\% = \frac{X}{425} \times 100$

## Section 3: MOLARITY

**1** Molarity is more accurate than percent in expressing concentrations. A *molar solution* contains 1 mole of solute per liter of solution. You will recall that a *mole* of a substance is its molecular weight in
grams ____________.

**2** To prepare a liter of molar sodium chloride (1 M NaCl), we first cal-
58.5 culate the molecular weight of the salt, which is ______. We then weigh out and dissolve this many grams of the compound in enough
1 water to make a volume of exactly ______ liter(s).

6 **3** To prepare 3 liters of a 2-molar NaOH solution, we need ______ moles.

240 **4** Six moles of NaOH equals ______ g.

## Section 4: NORMALITY AND EQUIVALENCE

**1** *Normality* is a useful way to express concentrations of acids and bases. A 1 normal solution contains 1 mole (1.008 g) of available hydrogen ion or its *equivalent* per liter. Thus a 6 N (six times normal)
6; 6.048 solution contains ______ moles, or ______ g of hydrogen ion per liter.

**2** When expressing the concentrations of bases, normality refers to available hydroxyl ion, $OH^-$. One $OH^-$ is the *equivalent* of 1 $H^+$ ion. Therefore, a 1 normal solution of a base contains 1 mole (17 g)
liter of available $OH^-$ per ____________.

**3** An *equivalent* is the amount of any substance that exactly combines with or replaces 1 mole of hydrogen. Thus *1* equivalent of $OH^-$ is
mole; 17 the same as *1* ____________ of $OH^-$ and equals ______ g.

**4** Although normality and equivalency are essentially the same for bases, we use normality to describe the concentration of laboratory solutions and equivalency to calculate stoichiometry of reactions.

**5** The normality of an acid that has 1 available hydrogen ion per molecule, such as HCl, is the same as its molarity. A solution of 36 g of
1 HCl in 1 liter of water is ______ molar. It will release 1 mole of $H^+$
1 on dissociation and therefore is ______ normal.

**6** $H_2SO_4$ contains 2 available $H^+$ per molecule. A 1 M solution of
2 $H_2SO_4$ will therefore be ______ normal.

**7** A simple rule of thumb is:

$$\text{normality} = \text{molarity} \times \text{number available } H^+ \text{ or } OH^- \text{ per molecule}$$

9 **8** The normality of a 3 M solution of $H_3PO_4$ is ______ N.

**9** To prepare a normal solution of $H_3PO_4$, divide

$$\frac{1 \text{ mole}}{3 \text{ available } H^+} = \frac{98}{3} = 33 \text{ g}$$

and add this amount of acid to enough water to make a volume
1 liter of ______ ________.

28 **10** To prepare 2 liters of a 0.4 N solution of $NH_4OH$, we need ______ g of $NH_4OH$.

*Calculation:*

$2 \times 0.4 = 0.8$ Eq
$35 \times 0.8 = 28$ g

2 **11** There is enough acid in 49 g of $H_2SO_4$ to make ______ liters of 0.5 N solution.

**12** Normality may also be applied to salts. Determine the equivalent weight of a salt, using the formula

$$\frac{\text{gram-molecular weight}}{\text{total positive or negative valence}} = \text{equivalent weight (Eq)}$$

Thus, for sodium chloride:

$$\frac{58.5}{1} = 58.5 \text{ g} = 1 \text{ Eq}$$

87 **13** The equivalent weight of $K_2SO_4$ is ______ g.

**14** The equivalent weight for individual ions may be derived by dividing the gram-atomic weight (atomic weight in grams) by the valence, or

$$\frac{\text{gram-atomic weight}}{\text{valence}} = \text{Eq}$$

20 The equivalent weight of $Ca^{++}$ is ______ g.

**15** A solution of 1 Eq of any salt in 1 liter volume yields a 1 normal solu-
85 tion. To prepare a 1 N solution of $NaNO_3$, ______ g $NaNO_3$ must be dissolved in sufficient water to make 1 liter.

171 **16** To make 5 liters of 0.3 N $Al_2(SO_4)_3$ we need ______ g $Al_2(SO_4)_3$.

*Calculation:*

$\text{Eq} = \frac{342}{3} = 114$
$5 \times 0.3 = 1.5$ Eq
$114 \times 1.5 = 171$ g

**17** In physiology, the *milliequivalent* (mEq) is used for the small amounts of dissolved salts in body fluids. This unit is derived as follows:

$$\frac{\text{gram-atomic weight}}{\text{valence} \times 1{,}000} = \text{mEq}$$

The mEq for the following are:

| | | | |
|---|---|---|---|
| 0.039 | $K^+$ | ______ | g |
| 0.020 | $Ca^{++}$ | ______ | g |
| 0.032 | $PO_4^{\equiv}$ | ______ | g |

**18** In normal blood there is a concentration of 140 mEq of $Na^+$ per liter.

3.22 This is equal to ______ g $Na^+$.

*Calculation:*

$$1 \text{ mEq} = \frac{23}{1 \times 1{,}000}$$

$$= 0.023 \text{ g}$$

$$140 \times 0.023 \text{ g} = 3.22 \text{ g}$$

## Section 5: TITRATION

**1** *Titration* is the laboratory procedure most frequently used to determine the concentration of an unknown solution. If we have a bottle marked only NaOH solution and we wish to know its *normality,* we would carry out a ______________.

titration

**2** Titration requires using a solution of *known* concentration, or a *standard solution,* to react with the unknown solution. In the case above, 1 N HCl solution would make a good ______________ ______________.

standard solution

**3** We would put an exact amount of the unknown NaOH solution in a beaker and slowly add the standard solution of HCl. As HCl is added to NaOH, ______________ occurs. The number of equivalents of HCl needed to completely neutralize the NaOH will equal the number of ______________ of NaOH present.

neutralization

equivalents

**4** We use an *indicator* to signal when we have added *just enough* HCl. The indicator, a chemical added in small amounts, signals this so-called *end point* by changing color. Phenolphthalein is a commonly used ______________ in titrations involving acids and bases. It is an *acid-base indicator,* because it changes from colorless in acid to magenta in base.

indicator

**5** In calculations based on titration, we use the formula

$$\text{milliliters acid} \times \text{normality of acid} = \text{milliliters base} \times \text{normality of base}$$

(that is, equivalents of acid equal equivalents of base). If we know three of these four values, the fourth is readily calculated. For example, if 36 ml of 1 N HCl is needed to neutralize 50 ml of unknown base, then X equals ______ N.

0.72

*Calculation:*

$$36 \times 1 = 50 \times X$$

**6** If it takes 30 ml of 0.2 N NaOH to neutralize 50 ml of $H_2SO_4$, then
0.12 the normality of the acid is ______ N.

*Calculation:*

50 X = 30(0.2)

**7** To titrate 32 ml of 0.15 *molar* $H_2SO_4$, the amount of 0.20 *molar*
48 NaOH needed is ______ ml NaOH.

*Calculation:*

0.20 X = 32(0.15 × 2)

## Section 6: pH AND BUFFERS

**1** The mathematical expression *pH* is used as a convenient quantitative measurement of the acid or base content of a solution. It is derived as follows.

**2** Pure water has been considered as a polar covalently bonded compound. However, $H_2O$ is *very slightly dissociated* and normally contains 0.0000001 mole of free $H^+$ (as $H_3O^+$) per liter. Because
0.0000001 water is *neutral,* we can predict that it will also contain ______ mole of $OH^-$.

**3** Numbers such as 0.0000001 are inconvenient to use, so they are converted to common logarithms, that is, logarithms to the base 10. The logarithm (log) of a number is the power or exponent to which 10 must be raised to equal that number. 0.0000001 converts exponentially to $1 \times 10^{-7}$, or simply $10^{-7}$, and the log of $10^{-7}$ is thus $-7$. (The log is *minus* because $10^{-7}$ is *less* than 1.)

**4** The expression $[H^+]$ means hydrogen-ion *concentration,* that is, the number of mole-ions per liter. The log of $[H^+]$ for pure water is
−7 ______.

**5** To avoid the use of *negative expressions,* we employ pH, which is defined as the *negative* logarithm of the $[H^+]$. Thus for water

$$[H^+] = 10^{-7}\ M$$
$$pH = -\log\ [H^+]$$
$$pH = -\log\ 10^{-7}$$
$$pH = \log\ 10^{7}\ \text{(because signs cancel)}$$
$$pH = 7$$

7 **6** A neutral solution, such as pure water, has a pH value of ______.

**7** For pure water we also define a constant, the *ion-product constant,* or (Kw):

$$Kw = [H^+] \times [OH^-]$$
$$Kw = [10^{-7}] \times [10^{-7}]$$
$$Kw = 10^{-14}\ \text{(exponents added in multiplication)}$$

**8** Kw is always $10^{-14}$; that is, if $[H^+]$ is $10^{-9}$, then $[OH^-]$ must be
$10^{-5}$ ______.

higher **9** An acid solution will have a *higher/lower* $[H^+]$ than water. Thus, a dilute HCl solution might have an $[H^+]$ of 0.01 M, or, exponentially,
$10^{-2}$; $10^{12}$ ______ M; the $[OH^-]$ would be ______ M.

2 **10** The pH of the above solution is ______.

lower

**11** An acid solution will have a pH value *lower/higher* than that of water.

**12** When we add a base to water, it dissociates to yield $OH^-$ ions. These react with free $H^+$ ions from the water as follows:

$$H^+ + OH^- \rightleftharpoons H_2O \qquad \text{(VII-1)}$$

less

Thus, there will be fewer free $H^+$ ions and the $[H^+]$ will be *less/greater* than $10^{-7}$ M.

12
increases
acceptor

**13** A base solution that reduces the concentration of free $H^+$ to $10^{-12}$ M will therefore have a pH of ________. The more $OH^-$ ions present, the fewer free $H^+$ ions. Therefore, the pH *increases/decreases* as available $OH^-$ or other proton ________________ is added.

**14** Calculating the pH of a solution whose $[H^+]$ is a decimal value, for example, $1.7 \times 10^{-3}$ M, requires use of the logarithm table at the end of this book. We proceed as follows:

$$\text{pH} = -(\log 1.7 + \log 10^{-3})$$
$$\log \text{ of } 1.7 = 0.2304 \text{ and}$$
$$\log \text{ of } 10^{-3} = -3$$
$$\text{so, pH} = -(0.2304 - 3)$$
$$\text{pH} = -(-2.77)$$
$$\text{pH} = \_\_\_\_\_\_$$

2.77

**15** The pH of a 0.0025 N solution of $H_2SO_4$ is:

0.0025
$10^{-3}$
0.3979
−2.6021
2.6

$$[H^+] = \_\_\_\_\_\_$$
$$\text{pH} = -(\log 2.5 + \log \_\_\_\_\_\_)$$
$$\text{pH} = -(\_\_\_\_\_\_ - 3)$$
$$\text{pH} = -(\_\_\_\_\_\_)$$
$$\text{pH} = \_\_\_\_\_\_$$

$10^{-14}$

**16** To calculate the pH of basic solutions, we convert $[OH^-]$ to exponential terms and divide this value into the Kw value, ________. Thus, the pH of a 0.01 N NaOH solution is

$$[OH^-] = 10^{-2}$$
$$10^{-14} \div 10^{-2} = 10^{-12} \text{ (exponents subtracted in division)}$$
$$\text{pH} = 12$$

**17** The pH of a 0.005 *molar* solution of $Ba(OH)_2$ is:

2
0.010
$10^{-2}$; $10^{-12}$
12

$$[OH^-] = 0.005 \times \_\_\_\_\_\_$$
$$[OH^-] = \_\_\_\_\_\_$$
$$10^{-14} \div \_\_\_\_\_\_ = \_\_\_\_\_\_$$
$$\text{pH} = \_\_\_\_\_\_$$

**18** A *buffer* is a solution that contains both weakly and strongly dissociated compounds and can correct small variations in pH to *maintain* a desired pH value. The maintenance of pH is called buffering. Blood and other body fluids resist changes in pH because they contain

buffers

________________.

**19** Let us consider a solution of $NH_4OH$ and $NH_4Cl$. These compounds dissociate as follows:

$$NH_4OH \rightleftharpoons NH_4^+ + OH^- \qquad \text{(VII-2)}$$
$$NH_4Cl \longrightarrow NH_4^+ + Cl^- \qquad \text{(VII-3)}$$

$NH_4^+$
$OH^+$; $Cl^-$
$NH_4OH$

The solution, then, contains three types of ions, ____________, ____________, and ____________. It also contains *undissociated* ____________. These entities exist in a state of equilibrium.

**20** When we add $H^+$ to the above solution, we disturb the equilibrium, because the $H^+$ will combine with the _____. However, the equilibrium is immediately reestablished by dissociation of additional _____. Thus, the pH will be *unchanged*. This buffering process will continue until all the $NH_4OH$ is _____, at which point further buffering cannot occur. No buffer has an unlimited capacity to maintain a desired _____.

$OH^-$
$NH_4OH$
dissociated
pH

**21** When we add $OH^-$, we disturb the equilibrium again. This time the added $OH^-$ combines with excess _____ to form *undissociated* $NH_4OH$. Again, the pH is _____.

$NH_4^+$
unchanged

**22** We can design a buffer system that can maintain pH at any selected value. Blood is an excellent buffer. Its pH is maintained at 7.4 by the interaction of a large number of ions in extremely complex equilibria.

## Section 7: BOILING POINT

**1** Each pure liquid has a definite boiling point, which is the temperature at which molecules pass from the liquid state to the gaseous state. For water this temperature is _____ °F, or _____ °C.

212; 100

**2** A liquid in a vessel loses some molecules to the space above it through *evaporation*. These gaseous molecules exert a *vapor pressure* against the surrounding atmosphere. When the liquid is heated, more molecules vaporize, and the _____ _____ increases.

vapor pressure

**3** When the vapor pressure *equals* the atmospheric pressure, the liquid boils. At high altitudes the atmospheric pressure is *lower* than at sea level. This means that water boils at a *higher/lower* temperature in the mountains.

lower

**4** When a *nonvolatile* (nonvaporizing) solute is dissolved in water, the solute molecules partly replace solvent molecules at the surface. Because fewer molecules of _____ will escape, the vapor pressure will be *higher/lower*.

solvent
lower

**5** To make the vapor pressure of the above solution equal to atmospheric pressure, *more/less* heat is required.

more

**6** Therefore, we would expect the boiling point of a solution (containing a nonvolatile solute) to be *higher/lower* than pure water.

higher

**7** For dilute solutions, the extent of the boiling point elevation is related to the *number* of particles in solution, not the kind of particle. This type of property is called a *colligative property*. A molecule of a nonelectrolyte contributes only 1 _____, whereas a molecule of an electrolyte contributes as many particles as it yields ions on dissociation. Thus, NaCl molecules yield _____ particles each.

particle
2

**8** The boiling point (bp) of aqueous solutions is given by the formula

$$\text{bp} = \text{moles solute per liter} \times \text{particles per molecule} \times 0.52°\text{C} + 100°\text{C}$$

Thus, if we add 1 mole of glucose, a nonelectrolyte, to a liter of water at sea level, the boiling point will be 100.52°C.

102.08

**9** If we add 2 moles of NaCl to 1 liter of water, the boiling point will be ________ °C.

*Calculation:*

bp = 2 × 2 × 0.52° + 100°

## Section 8: FREEZING POINT

**1** Each pure liquid exhibits a characteristic *freezing point* (fp) which is the point at which its molecules organize into a rigid three-dimensional structure and acquire a fixed shape. Freezing points are *depressed* by the addition of solutes. Recall that freshwater freezes at a *higher/lower* temperature than saltwater.

higher

**2** Freezing point depression is also dependent only on the number of particles in solution. Like boiling point elevation, it is a __________ property.

colligative

**3** For aqueous solutions, the freezing point is found by using the formula

$$\text{fp} = 0°\text{C} - (\text{moles solute per liter} \times \text{number particles} \times 1.86°\text{C})$$

Thus, if we add 1 mole of glucose to 1 liter of water, its freezing point will be __________.

−1.86°C

*Calculation:*

fp = 0°C − (1 × 1 × 1.86°C)

**4** If 2 moles of $K_2SO_4$ are added to 1 liter of water, its freezing point will be __________.

−11.16°C

*Calculation:*

fp = 0°C − (2 × 3 × 1.86°C)

## Section 9: OSMOSIS

**1** Osmosis is a physical process involving the passage of water or other solvent molecules through a *semipermeable membrane* separating two solutions. Such a membrane is a thin, flexible film of natural or artificial substance. It is said to be semipermeable because it permits the passage of *solvent* molecules but inhibits or prevents the passage of *solute* __________.

molecules

**2** Water molecules can pass in both directions through a membrane. If one solution contains more dissolved solute than the other, water (or solvent) will flow into the *more* concentrated solution. In other words, water will move into the solution with the *higher/lower* number of dissolved particles.

higher

**3** Because the direction of osmotic flow is determined only by the *number of dissolved particles,* we are dealing once again with a(n) __________ property.

colligative

**4** Osmosis continues between two solutions of unequal solute particle concentration until both solutions attain *equal* concentrations. At this point, the number of water molecules passing each way is *equal*. In other words, we have osmotic ________________.

equilibrium

**5** The flow, or force, with which water molecules pass into the more concentrated solution can be measured by determining the force necessary to *prevent* the movement. This preventative force is called *osmotic pressure*. It will prevent solvent movement if applied to the solution of greater ____________ particle concentration.

solute

**6** The higher the concentration of solute, the larger the magnitude of the osmotic pressure required to prevent solvent movement. Based on what has been said previously, a 1 molar glucose solution will have a *higher/lower* osmotic pressure than a 1 molar NaCl solution.

lower

**7** A living cell contains *cytoplasm*, a watery solution of many dissolved materials surrounded by a semipermeable membrane. Most cells, when placed in pure water, will swell because water enters the cell as a consequence of the process of ________________.

osmosis

**8** This swelling may result in the rupture (plasmoptysis) and death of the cell. When the exterior solution is more dilute than the interior solution, it is said to be *hypotonic*.

**9** When a cell is placed in a solution *more* concentrated than its cytoplasm, it will lose water and shrink. Such a solution is said to be *hypertonic*, and the shrinking is referred to as *plasmolysis*.

**10** Thus, hypotonic solutions cause ________________, whereas ________________ solutions cause plasmolysis.

swelling
hypertonic

**11** A solution with an osmotic pressure equal to that of a given cell is said to be *isotonic*.

**12** Because blood plasma maintains blood cells at their normal size and shape, it must be ________________.

isotonic

**13** Isotonic, hypertonic, and ________________ are *relative* terms.

hypotonic

PART **VIII**

# Colloids

## Section 1: COLLOIDAL STATE

**1** We have discussed solutions in which the added solutes, sometimes called *crystalloids,* dissolve completely and form clear homogeneous mixtures.

**2** Suspensions are mixtures that are neither clear nor stable. If we add sand to water and stir, we get a *suspension*. Particles in suspension settle out.

suspension — **3** The *colloidal state* lies between solutions and suspensions. In this state, the particles are *larger* than those of a solution, but *smaller* than those of a(n) ________________.

colloidal — **4** The term colloid is usually used to characterize the ________________ state.

**5** Characterize the following as solutions, colloids, or suspensions:

| | | |
|---|---|---|
| solution | Tea | ________________ |
| colloid | Milk | ________________ |
| colloid | India ink | ________________ |
| suspension | Muddy water | ________________ |

solutions — **6** Many biologically important fluids, such as blood, milk, and cellular cytoplasm, are colloids. It is important to realize that particle size is the key difference between colloids, suspensions, and ________. Large clay particles form suspensions, small ones form colloids, and the compounds that make up clay form solutions as single molecules.

**7** Colloids may consist of solids, liquids, or gases *dispersed* in solids, liquids, or gases. Solids dispersed in liquids are called *sols* if the resulting colloids are fluids and *gels* if they are firm. Liquids dispersed in liquids are called *emulsions*. Classify the following types of colloids:

| | | |
|---|---|---|
| emulsion | Milk | ________________ |
| sol | Paint | ________________ |
| gel | Jell-O | ________________ |

solutions — **8** *Emulsions* are formed from *immiscible* liquids. Miscible liquids form true ________________.

**9** Oil and vinegar shaken together form an unstable emulsion that readily separates. However, shaking the mixture with egg yolk emulsion — produces mayonnaise, a stable ________________. Egg yolk is called the *emulsifying agent*.

emulsifying agent — **10** Similarly, to wash oil and grease from dishes, we use soap or detergent as the ________________ ________________. These agents keep the oil and grease particles from forming droplets that would exceed colloidal size and thereby separate from the water.

## Section 2: PROPERTIES OF COLLOIDS

brownian — **1** All particles of matter exhibit motion. This is true of atoms, molecules, and colloidal particles. When we observe colloids under a microscope, the motion of the dispersed particles is visible. This movement is called *brownian movement*. Colloidal particles are moved about as a result of collisions with molecules of the dispersion medium. ________________ movement helps to keep colloidal particles from settling.

solution — **2** When a beam of light shines through a colloid, its path is visible, as with a ray of sun passing through a smoky room. This is called the *Tyndall effect*. Colloidal particles are large enough to scatter and reflect light, whereas, particles of a(n) ________________ are too small to do so.

repel — **3** Colloidal particles usually have an electrical charge, and all particles in a given system have the *same* charge. Dispersion of the particles is assisted by the fact that like charges *repel/attract* each other.

charged — **4** We can destroy the stability of a colloid by adding an electrolyte as a source of opposite charges. A pinch of salt clarifies muddy coffee, because the ions neutralize the ________________ particles and cause them to aggregate and settle out.

neutralized — **5** When a direct current is passed through a colloid, the dispersed particles travel to the oppositely charged electrode. There they are ________________ and precipitate out. This process is called *electrostatic precipitation*. It is used for cleaning up emissions of industrial smoke stacks.

**6** A related process used in biochemistry, called electrophoresis, is used to separate proteins from blood plasma and certain other colloidal mixtures.

## Section 3: DIALYSIS

**1** *Dialysis* is a process that involves selective diffusion through a special *semipermeable* membrane separating a colloid and pure water or a solution.

membranes — **2** Some *semipermeable* ________________ restrain colloids but pass crystalloids. The membranes of our kidneys have this ability.

**3** If we place a mixture of water, sugar (a crystalloid), and protein (a colloid) on one side of a semipermeable membrane and plain water sugar on the other, only the water and __________ will pass through the membrane and leave the mixture.

4 Crystalloids pass through the membrane until their concentrations

equal are __________ on both sides. Water will also flow to equalize the

osmotic __________ pressure on both sides.

5 A common medical application of dialysis is the artificial kidney, through which blood is dialyzed against a sterile solution containing desirable crystalloids at the proper concentrations. Neither blood cells nor colloids (proteins) can pass through the semipermeable

membrane __________, but crystalloids do; thus, essential crystalloids pass into and waste crystalloids pass out of the blood.

6 In the artificial kidney, crystalloids diffuse through a semipermeable membrane in both directions to establish balance, or equilibrium, the direction of the diffusion depending on the concentration.

waste __________ crystalloids diffuse into the sterile dialyzing solution.

PART **IX**

# Nuclear chemistry

## Section 1: RADIOACTIVITY

**1** In the early part of this century it was discovered that certain natural minerals give off invisible radiations that cause chemical reactions similar to light energy; that is, they expose photographic film, cause fluorescence, and so forth. This property is called *radioactivity*. Materials that emit radiation are said to be ______________.

radioactive

**2** The first sources of radiation were naturally occurring isotopes. You will recall that isotopes are different forms of the same atom or element, with different numbers of *neutrons*. All isotopes of an element, however, have the same number of ______________ and ______________.

protons
electrons

**3** Lead, for example, has several isotopes. All have the same atomic number, 82. However, they have different atomic weights, more correctly called mass numbers when we speak of the mass of the specific isotope. Thus we can symbolize them as follows:

$$^{214}_{82}Pb,\ ^{210}_{82}Pb,\ ^{206}_{82}Pb$$

These 3 lead isotopes have different mass numbers (superscript number) because they have different numbers of ______________. They all have 82 positive charges, or protons, in their nucleus, as shown by the subscript number.

neutrons

**4** Many isotopes are *unstable*. That is, they tend to undergo *nuclear reactions* that convert them to different isotopes or new elements. Because these conversions result in the emission of radiation, unstable isotopes are said to be ______________.

radioactive

**5** Radioactive isotopes undergo disintegration, or decay, at a definite rate. We define this rate in terms of *half-life*. The half-life of an isotope is the length of time necessary for one half of a sample of the isotope to be converted to its more stable form. Thus radium, which has a half-life of 1,620 years, will be half converted in that time. If we have 1,000 units of radioactive radium now, in 1,620 years we will have only ________ units.

500

**6** In the second 1,620 years, half *again* will decay, so we will have ________ units left.

250

**7** Other radioactive isotopes have half-lives measured in minutes or fractions of seconds.

**8** Radioactive isotopes give off three types of radiation (rays): alpha ($\alpha$), beta ($\beta$), and gamma ($\gamma$). Alpha and beta rays are *particulate* in nature; that is, they are composed of streams of particles. Alpha particles are equivalent to helium nuclei. Referring to the

4 periodic table, we see that helium has an atomic weight of ______

2 and an atomic number of ______. The alpha particle, in nuclear notation, is written

$$\begin{matrix}\text{mass number} \\ \text{protons}\end{matrix}\ {}^{4}_{2}\alpha$$

electrons Because it is a nucleus, it has no ______.

**9** Beta rays are streams of electrons of negligible mass. They are shown as

$${}^{0}_{-1}\beta$$

negative They have 1 *positive/negative* electrical charge, shown by the subscript number.

**10** Gamma rays are not particulate, but electromagnetic radiations such as light rays and x-rays. Because this form of radiation is a form of

mass (weight) energy, it has no ______.

**11** Radioactivity, then, is the emission of radiation by unstable

isotopes; $\alpha$; $\beta$ ______ in one of three forms, ______ or ______

$\gamma$ particles or ______ rays.

## Section 2: NUCLEAR REACTIONS

**1** Let us learn how to write balanced equations for nuclear reactions. First, look at a conventional equation:

$$HCl + NaOH \longrightarrow NaCl + H_2O \qquad \text{(IX-1)}$$

All atoms present participate and emerge unchanged. In conventional reactions only one type of subatomic particle is involved, the

electron ______.

**2** In contrast, look at a nuclear reaction:

$${}^{238}_{92}U \longrightarrow {}^{4}_{2}\alpha + {}^{234}_{90}Th \qquad \text{(IX-2)}$$

**3** This equation reads: uranium (U) loses an alpha particle and becomes thorium (Th). The equation illustrates *transmutation*. Uranium and thorium are different elements because they have different numbers of protons (atomic numbers). When one element is converted to

transmutation another, the process is called ______.

**4** In balancing a nuclear reaction, we must balance the *masses* on either

mass side. Thus, in equation IX-2 we have 238 units of ______ on each side.

**5** Balance this equation by adding the missing mass and atomic numbers:

${}^{4}_{2}\alpha$; ${}^{210}_{81}Tl$

$${}^{214}_{83}Bi \longrightarrow {}^{\_}_{\_}\alpha + {}^{210}_{\_}Tl \qquad \text{(IX-3)}$$

electron

**6** To balance equations in which $\beta$ particles are given off, we must realize that a $\beta$ particle, or ________________, arises from the splitting of a neutron. Neutrons are electrically neutral and have 1 mass unit. When they split, they yield 1 proton and 1 electron:

$$\underset{\text{(neutral)}}{{}^{1}_{0}\text{n}} \longrightarrow \underset{\text{(positive)}}{{}^{1}_{1}\text{p}} + \underset{\text{(negative)}}{{}^{0}_{-1}\text{e}} \qquad \text{(IX-4)}$$

*Superscript* numbers equal *masses; subscript* numbers equal *charges*. Both sets of numbers *balance*.

**7** For example, look at a nuclear reaction where a $\beta$ particle is emitted:

$$^{239}_{92}\text{U} \longrightarrow {}^{239}_{93}\text{Np} + {}^{0}_{-1}\beta \qquad \text{(IX-5)}$$

93

Note that the number of positive charges or protons increases from 92 to ________. Thus we have a new element, neptunium (Np). This reaction is also a transmutation. In equation IX-5, we see that both masses and charges balance.

**8** Balance this equation by filling in any missing numbers:

$^{214}_{84}\text{Po}$; ${}^{0}_{-1}\beta$

$$^{214}_{83}\text{Bi} \longrightarrow {}^{214}_{\_\_}\text{Po} + {}^{\_\_}_{\_\_}\beta \qquad \text{(IX-6)}$$

**9** The particle given off in this reaction is:

${}^{4}_{2}\alpha$

$$^{230}_{90}\text{Th} \longrightarrow {}^{226}_{88}\text{Ra} + {}^{\_\_}_{\_\_}\alpha \qquad \text{(IX-7)}$$

**10** The mass number of the polonium (Po) isotope in this equation is:

$^{210}_{84}\text{Po}$

$$^{\_\_}_{84}\text{Po} \longrightarrow {}^{206}_{82}\text{Pb} + {}^{4}_{2}\alpha \qquad \text{(IX-8)}$$

## Section 3: NUCLEAR FISSION

**1** Certain nuclear reactions may emit *neutrons* as a fourth type of radiation. When a stable atomic nucleus is hit by such a neutron, it may become unstable and actually split. This is called *nuclear fission,* the products of which include 2 new nuclei, energy, and more neutrons.

**2** For example:

$$^{235}_{92}\text{U} + {}^{1}_{0}\text{n} \longrightarrow {}^{92}_{36}\text{Kr} + {}^{141}_{56}\text{Ba} + 3\,{}^{1}_{0}\text{n} + \text{energy} \qquad \text{(IX-9)}$$

Note that 3 *neutrons* are released.

**3** The neutron bombardment of uranium 235 yields rubidium (Rb) and cesium (Cs). The balanced equation is:

$^{90}_{37}\text{Rb}$; $^{143}_{55}\text{Cs}$; $^{1}_{0}\text{n}$

$$^{235}_{92}\text{U} + {}^{1}_{0}\text{n} \longrightarrow {}^{90}_{\_\_}\text{Rb} + {}^{\_\_}_{55}\text{Cs} + 3\,{}^{\_\_}_{\_\_}\text{n} + \text{energy} \qquad \text{(IX-10)}$$

fission

3

9

**4** Because each of the 3 neutrons released in the above reaction can trigger another nuclear ________________, in the next round of reactions (under the appropriate conditions) ________ atoms of uranium will split and yield a total of ________ neutrons.

27

**5** These 9 neutrons released will in turn fission the release of 9 × 3, or ________, more neutrons, and so on. This is a *chain reaction:* the energy released is *multiplied* at each step and rapidly reaches the chain

level of an explosion. Thus, the *atomic bomb* is an *uncontrolled* ____________ reaction that releases huge amounts of energy in a short time.

energy

**6** Chain reactions can be controlled, as evidenced by nuclear power plants that generate electricity. In controlled fission, ____________ is released gradually.

1

**7** To control fission we can introduce some material into the reactor that absorbs neutrons efficiently *without* undergoing fission itself. Thus, of the 3 neutrons released per fission, 2 may be absorbed, leaving ______ neutron(s) to *perpetuate* the energy-releasing reaction.

neutrons

**8** *Control rods* and *moderators* of cadmium and boron compounds are the *nonfissionable* materials commonly used in atomic power plants to absorb excess ________________.

## Section 4: MAN-MADE ISOTOPES

**1** In addition to the natural radioactive isotopes, there are man-made isotopes produced by bombarding nonradioactive isotopes with neutrons.

**2** For example, the useful medical isotope cobalt 60 is produced by the reaction:

$$^{59}_{27}Co + ^{1}_{0}n \longrightarrow ^{60}_{27}Co \qquad \text{(IX-11)}$$

$^{14}_{6}C$; $^{131}_{53}I$; $^{38}_{19}K$

**3** Other man-made isotopes include:

$$_{\_}^{14}C,\ _{\_}^{131}I, \text{ and } _{\_}^{38}K$$

radioactive

**4** Nuclear fission, controlled or uncontrolled, produces many nuclear waste products. Many of these are unstable and thus ____________. These *nuclear wastes* are hazardous to all forms of life. For example, fission of uranium yields radioactive strontium 90, or

$$^{90}_{38}Sr$$

which the body metabolizes like calcium. Once in bone, this isotope undergoes decay and in so doing releases beta rays:

$^{0}_{-1}\beta$

$$^{90}_{38}Sr \longrightarrow ^{90}_{39}Y + {}_{\_}^{\_}\beta \qquad \text{(IX-12)}$$

Further,

$^{90}_{40}Zr$; $^{0}_{-1}\beta$

$$^{90}_{39}Y \longrightarrow {}_{\_}^{90}Zr + {}_{\_}^{\_}\beta \qquad \text{(IX-13)}$$

These $\beta$ rays are harmful to bone marrow and other tissues, sometimes causing anemia and cancer.

## Section 5: NUCLEAR ENERGY

created

**1** When atomic nuclei split, energy is released. Where does this energy come from? In Part I we stated that under ordinary circumstances, energy and matter are conserved; that is, they are neither ________________ nor destroyed.

2 Einstein predicted that under extraordinary conditions, mass (m) and energy (e) could be interconverted. The relativity equation

$$E = mc^2$$

relates this conversion to the speed of light (c). Thus, matter can be converted to ________ and thereby ________.

energy; destroyed

3 In reverse, energy can be converted to ________. In so doing, energy is ________ and matter is ________.

matter

destroyed; created

4 When we add up the masses of all the *individual* protons and neutrons in an atomic nucleus, we find that they weigh slightly more than the nucleus itself. This tells us that in assembling the nucleus some matter has been ________ or converted to ________.

destroyed; energy

5 This loss of mass is called the *mass defect*. The energy created holds the nucleus together; it is called *binding energy*. We can say that when a nucleus is assembled, matter is destroyed (the ________ ________) and ________ energy is created.

mass defect; binding

6 Small nuclei require much less binding energy than large nuclei. When fission occurs, small nuclei are produced from large ones. There is, then, a substantial amount of energy released in the fission of 1 atomic nucleus. Complete conversion of 1 g of matter to energy would yield $9 \times 10^{13}$ joules of energy.

## Section 6: NUCLEAR FUSION

1 Nuclei can also *fuse* and give off energy. For example, the following is one of the possible reactions in the *hydrogen bomb:*

$$^{7}_{3}Li + ^{1}_{1}H \longrightarrow 2\ ^{4}_{2}He + \text{energy} \qquad (IX\text{-}14)$$

2 *Fusion* reactions take place among small nuclei. As small nuclei fuse, a *mass defect* appears; that is, some ________ is converted into energy. Most of the energy produced is released because all of it is not required to bind the nuclei together.

mass

3 Balance these fusion reactions:

$$^{3}_{1}H + ^{1}_{1}H \longrightarrow \underline{\ \ }He \qquad (IX\text{-}15)$$

$^{4}_{2}He$

$$^{2}_{1}H + ^{3}_{1}H \longrightarrow \underline{\ \ }He + \underline{\qquad}\ \text{(particle)} \qquad (IX\text{-}16)$$

$^{4}_{2}He + ^{1}_{0}n$

4 Controlled nuclear fusion reactions as a source of energy are not yet technically possible, because tremendous heat and pressure are required to induce fusion. We on the earth, however, benefit from solar energy, which is produced as a result of ________ fusion.

nuclear

## Section 7: EFFECTS OF RADIATION ON MATTER

1 We have already seen that certain forms of radiation can cause atomic ________ to split and form new radioactive *isotopes*. Certain other forms can expose photographic film, an effect used in making x-ray photographs. X-radiation is a type of man-made electromagnetic radiation.

nuclei

**2** Different sorts of radiation penetrate matter to various depths: $\alpha$ particles have little penetrating power, $\beta$ particles have more, and $\gamma$-rays have the most. In fact, when we make x-ray films, we take advantage of the fact that the radiation penetrates through soft tissues but not bone. Thus, bone casts "shadows" by preventing the rays from exposing the film.

**3** Radiation may cause atoms of various elements to *ionize,* an effect that can be detected, measured, and recorded by a Geiger counter. Free ions in living matter can cause unexpected reactions, many of which are damaging and sometimes lethal. The biological effects of radiation are due in great part to its ability to ______________ cellular elements.

ionize

**4** Tissue subjected to ionizing radiation may die. This is why we irradiate cancerous tissues. However, under other conditions irradiated cells may undergo genetic change and become cancerous. Fortunately, we can control both processes to some extent. Biologically, the risks of radiation include tissue ______________ and genetic ______________. Its benefits include a wide range of diagnostic procedures *(roentgenography, fluoroscopy, scanning)* and radiotherapy (for example, radiation of ______________ tissue).

death
change
cancerous

PART **X**

# Organic chemistry: the hydrocarbons

## Section 1: STRUCTURAL FORMULAS

**1** *Organic chemistry* is the study of carbon compounds, with the exception of the gases CO and $CO_2$ and carbonate salts. This includes practically every chemical entity found in living organisms.

**2** *Hydrocarbons* are the simplest of this vast group of chemicals. They are compounds of carbon and hydrogen. Although this restricts us to

*carbon* combinations of only two elements, we will show that __________

*hydrogen* and __________ combine in such a way as to produce a vast number of compounds.

**3** Much of the variety in organic compounds comes not from the numbers and types of atoms specified by the *empirical* formula, but from their *arrangement*. Thus, it is necessary to use drawings of molecules, or *structural* formulas, to describe organic compounds.

**4** For example, methane, the simplest hydrocarbon, can be designated as:

*empirical* $CH_4$ (an __________ formula)

or

```
    H
    |
H—C—H (a __________ formula)      (I)
    |
    H
```

*structural*

**Methane**

**5** Remember that carbon forms covalent bonds with hydrogen. When these two elements combine to form methane or any other

*hydrocarbon* __________, they must obey the valence rules. Carbon has

*4; 1* a valence of ______, and hydrogen a valence of ______.

**6** Thus, the four single lines joining the carbon atom of methane to the 4

*covalent* hydrogen atoms represent four __________ bonds, or four

*electrons* pairs of shared __________.

## Section 2: ALKANES

**1** *Alkanes* (Table 6) are the simplest hydrocarbons. Their empirical formulas are derived from the *generic*, or group, formula $C_nH_{2n+2}$.

**Table 6.** Alkanes: the paraffin series

| *Formula* | *Name* | *Number of isomers* |
|---|---|---|
| $CH_4$ | Methane | — |
| $C_2H_6$ | Ethane | — |
| $C_3H_8$ | Propane | — |
| $C_4H_{10}$ | Butane | 2 |
| $C_5H_{12}$ | Pentane | 3 |
| $C_6H_{14}$ | Hexane | 5 |
| $C_7H_{16}$ | Heptane | 9 |
| $C_8H_{18}$ | Octane | 18 |
| $C_9H_{20}$ | Nonane | 35 |
| $C_{10}H_{22}$ | Decane | 75 |

Methane is the simplest alkane. Notice that $CH_4 = C_{(1)}H_{2(1)+2}$. Applying the generic formula, write the empirical formula for ethane, the alkane with 2 carbon atoms:

$C_2H_6$

C_H_

**2** In drawing the structural formula for ethane ($C_2H_6$), we see that carbon, valence 4, bonds not only to hydrogen but also to another carbon atom. Hydrogen, valence 1, can bond only to 1 ________ atom in a hydrocarbon molecule.

carbon

**3** To fit together 2 carbons and 6 hydrogens without breaking any valence laws, we can draw only one structure:

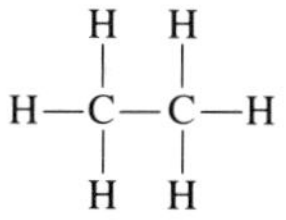

(II)

**Ethane**

Satisfying yourself that this is the only possibility, draw the structural formula for propane, the 3-carbon alkane, $C_3H$ __:

8

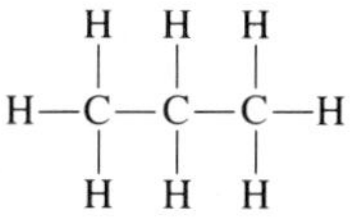

**Propane**

(III)

**4** For the names of the alkanes, refer to Table 6. You will see that the number of carbons determines the *prefix;* all have the suffix *-ane.* Thus, the 5-carbon alkane is called ________________, and the 8-carbon alkane, ________________. Memorize these ten prefixes, because they will turn up repeatedly in naming organic compounds.

pentane

octane

**5** Let us consider the alkane butane, $C_4H_{10}$. There are two ways to draw the structure for this molecule that are consistent with the rules of valence. These are shown on p. 62.

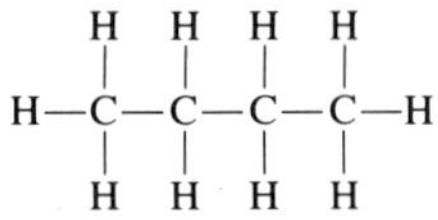

(IV)

**Normal butane**

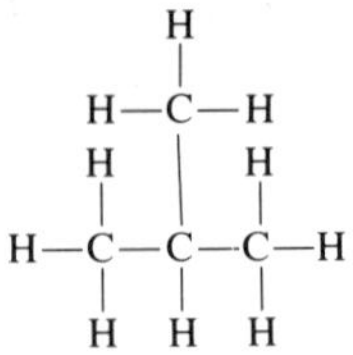

(V)

**Isobutane**

Compounds such as these, which have the same *empirical* formula but different *structural* formulas, are called *isomers*. Thus, we can
isomers — say that there are 2 ________________ of butane. Referring to Table 6, we see that as the alkanes increase in molecular size, the number of structural possibilities increases dramatically. Thus, it is
75 — possible for decane to have ________ isomers.

**6** For naming these isomers, organic chemists follow the rules set down by the International Union of Pure and Applied Chemistry (IUPAC).

**7** Structures IV and V above show *chains* of carbons. All such compounds are said to be *aliphatic*. Structure IV is a *straight-chain*, or *normal*, aliphatic compound, whereas structure V is a *branched-chain* aliphatic compound.

aliphatic — **8** All the compounds in Table 6 and their isomers are ________________ compounds. For each member of the series beyond propane, there will be a certain number of possible isomers. There can be just one
normal — straight-chain, or ________________, isomer; the others are
branched — ________________-chain isomers.

## Section 3: ALKENES

**1** *Alkenes* are the next class of aliphatic hydrocarbons. Their generic formula is $C_nH_{2n}$. They are named by combining the same prefixes used for alkanes, with the suffix *-ene*. Thus, the 2-carbon alkene is
heptene — *ethene*, and the 7-carbon alkene is ________________.
4 — **2** In trying to draw ethene, which has the empirical formula $C_2H\_\_$, we find that the only permissible structure is one with a carbon-to-carbon double bond:

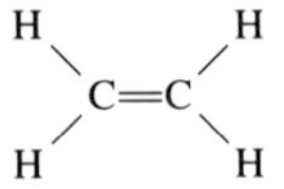

(VI)

**Ethene**

All alkenes contain at least 1 *double* covalent bond somewhere along the chain. The two possible structures for butene ($C_4H_8$) are:

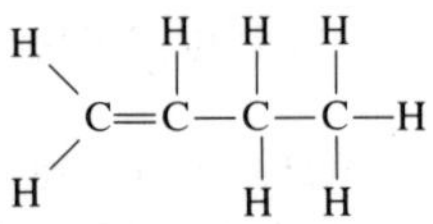

(VII)

or

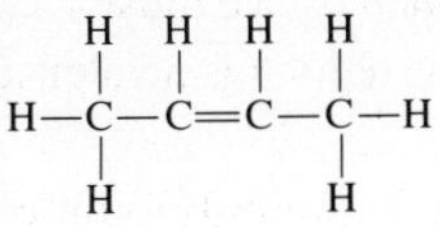

**Butene**

(VIII)

hydrogen

saturated

3 The basic difference between alkenes and alkanes is that alkenes contain less hydrogen per carbon than alkanes. Because they contain less than the maximum amount of ________________, alkenes are said to be *unsaturated*. Alkanes, by contrast, must be ________________ with hydrogen.

## Section 4: ALKYNES

octyne

1 Alkynes are aliphatic hydrocarbons with the generic formula $C_nH_{2n-2}$. Their names are composed from the appropriate prefix plus the suffix *-yne*. Thus, the 2-carbon alkyne has the name ethyne, and the 8-carbon alkyne ________________.

2 To draw structural formulas for the alkynes we must use a triple bond. For example, the structure for ethyne ($C_2H_2$), commonly called acetylene, is:

$$H-C\equiv C-H \qquad \text{(IX)}$$

**Ethyne**

4

Draw a structural formula for propyne, $C_3H$__.

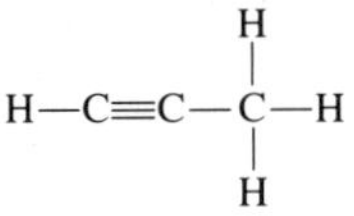

**Propyne**

(X)

ethyne

3 The stability of triple covalent bonds is very low, double bonds somewhat higher, and single bonds very high. Thus, if we compare ethyne ($C_2H_2$), ethene ($C_2H_4$), and ethane ($C_2H_6$), we can predict that ethane will be the most stable molecule. The least stable will be ________________.

## Section 5: CYCLIC HYDROCARBONS

carbocyclic

heterocyclic; cyclic

1 *Cyclic* compounds, as their name implies, contain *ring* structures. They are subdivided into those in which the ring is *all* carbon, *carbocyclic* compounds, and those in which 1 or more nitrogen, sulfur, or oxygen atoms are found in the ring, *heterocyclic* compounds. Thus, structure XI is ________________, and structure XII is ________________. Both are ____________ compounds.

(XI)

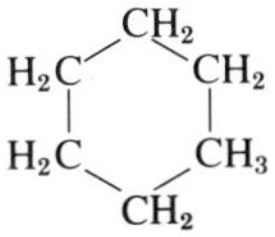

**Cyclohexane**

(XII)

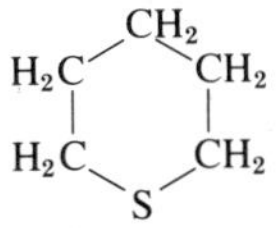

**Pentamethylene sulfide**

2 — **2** In drawing the structures for heterocyclic compounds, we must keep in mind that, when bonded to carbon, nitrogen usually has a valence of 3; sulfur, 2. Oxygen has a valence of ______.

valences — **3** In drawing structural formulas, hydrogens are commonly omitted for the sake of convenience; that is, we assume hydrogens to be present to *satisfy* all necessary ______________. In the case of cyclic compounds we employ the appropriate ring symbols. For example, structure XI may be written:

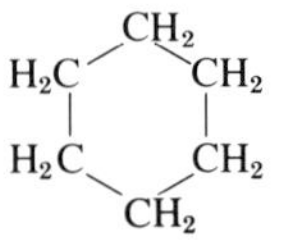

(XIII)

or

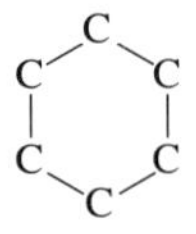

(XIV)

**Cyclohexane**

or

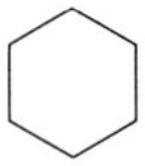

(XV)

hydrogens — In structure XV, each corner represents *1* carbon with *2* ______________ bonded to it.

**4** If the compound is *heterocyclic,* it is equally important to specify the location and valence of the noncarbon atoms in the ring. Thus structure XII becomes:

(XVI)

**5** Cyclic compounds may contain double bonds, too; for example:

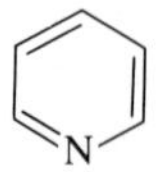

(XVII)

heterocyclic — Further, because the ring contains an atom of nitrogen, this compound is ______________.

## Section 6: AROMATIC COMPOUNDS

**1** A special class of cyclic hydrocarbons consists of six-membered carbocyclic compounds containing *3* double bonds. These compounds are commonly referred to as *aromatic* hydrocarbons, the simplest of which is benzene ($C_6H_6$), with the structure:

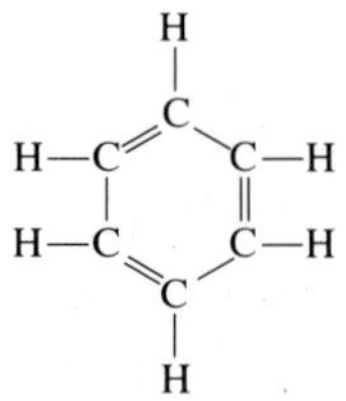

(XVIII)

double — **2** Note that the 3 ______________ bonds are drawn in alternate positions. However, these bonds are not really fixed in place. Where

single — double bonds and ______________ bonds alternate, a molecular flip-flop, called *resonance,* can occur, as it does in this case:

(XIX) (XX) (XXI)

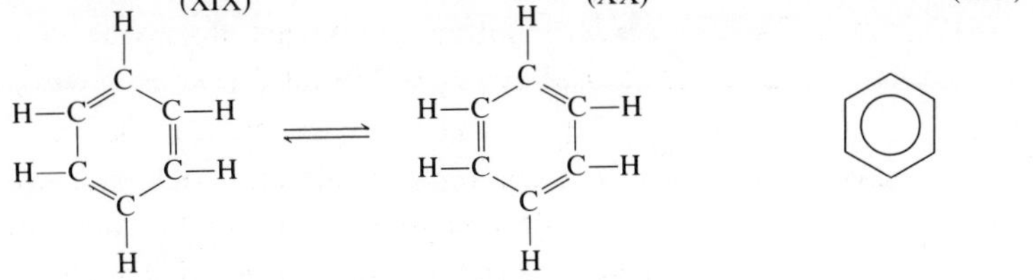

Structures XIX and XX are fully equivalent resonance forms. Both are adequately represented by XXI, in which the inner circle symbolizes the 3 migrating ________ ________.

double bonds

**3** The aromatic hydrocarbons are very reactive because of the mobility of their bonds, that is, because of their ________.

resonance

**4** Many dyes, drugs, disinfectants, and other important commercial organic products are aromatic derivatives.

## Section 7: ADDITION POLYMERS

**1** Each of the groups of hydrocarbons illustrates the principle of the carbon-to-carbon bond. There is no theoretical limit to the length of carbon chains, or backbones. We have taken advantage of this potential in developing the giant molecules of plastics and related materials. Plastics are manmade chemicals with exceedingly long carbon ________.

backbones (chains)

**2** Plastics are members of a class of compounds called polymers. They are manufactured by causing many smaller molecules to link together. The small subunits, or *monomers,* are joined to form the giant molecules, or ________. This process, called *polymerization,* is of as much importance to biochemistry as to industrial chemistry.

polymers

**3** Polymers formed from hydrocarbon ________ are joined by *addition* reactions. They are therefore called ________ polymers.

monomers
addition

**4** Let us look at the formation of polyethylene, a familiar plastic. Its monomer is ethene ($C_2H_4$), or, as commonly called, ethylene. Formation of the plastic involves addition of 1 ethene molecule to the next:

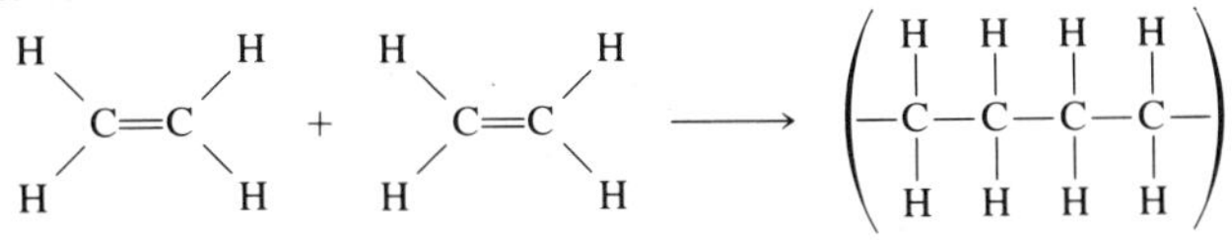

(X-1)

There is no limit to length. If 100 ethene molecules are used up in the reaction, the resulting polymer will be ________ carbon atoms long.

200

**5** Notice that as two monomers link together, the link is formed by replacing the double bonds with single bonds. Thus we clearly see that *addition* polymerization simply involves the ________ of each monomer unit to the next. Most of the important manmade polymers are of this type.

addition

PART **XI**

# Organic chemistry: substituted hydrocarbons

## Section 1: HALOGENATED HYDROCARBONS

**1** The simple hydrocarbons, described in Part X, are a large and important group of compounds found in petroleum and are indispensible to industry. *Substituted hydrocarbons,* compounds containing some element or elements *other* than carbon and hydrogen, include those essential to life. Substituted hydrocarbons contain C, H, and some other ________________.

element

**2** A halogenated hydrocarbon, we can suppose, will contain an atom or atoms of F, Cl, Br, or ______.

I

**3** Let us look at a very simple example, $CH_3Cl$:

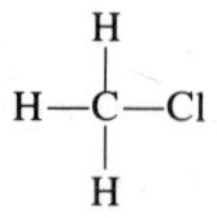

(I)

This compound is *methyl chloride*. It is a substituted form of $CH_4$, or ________________.

methane

**4** The component ($CH_3$—), if separated from its fourth bonding partner by breaking a covalent bond, would have an *unpaired electron,* shown as

$$CH_3\cdot \text{ and } \cdot Cl \qquad \text{(II)}$$

This makes it an entity called a *free radical*.

**5** We name alkane *radicals* by taking the name of the parent hydrocarbon and changing the *-ane* suffix to *-yl*. Thus, methane yields methyl, butane yields ____________, and ____________ yields octyl radicals.

butyl; octane

**6** Organic radicals do not have electrical charges as do such inorganic radicals as $NO_3^-$, $NH_4^+$, and others you are familiar with, but they are very reactive chemically. Because organic radicals are not charged, they react to form *covalent/electrovalent* compounds.

covalent

**7** All of the substituted hydrocarbons contain an organic radical. Thus, $CH_3Cl$ consists of a methyl ________________ in combination with a halide.

radical

**8** Halogenated hydrocarbons are very rare in nature. Most are man-made by either *substitution*

$$CH_4 + Cl_2 \longrightarrow CH_3Cl + HCl \qquad \text{(XI-1)}$$

or addition

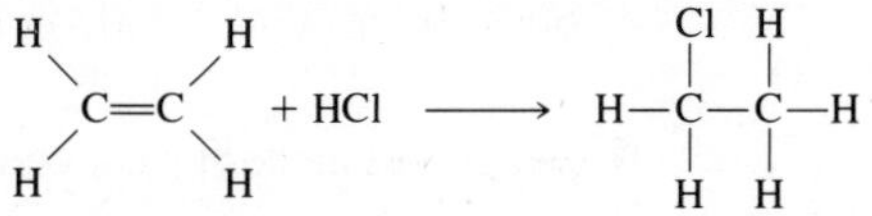

(XI-2)

Thus, a chemist may convert a simple hydrocarbon to a halogen derivative by causing a halogen to replace a ________ or by adding 2 atoms, one or both being halogens, to a ________ bond.

hydrogen
double

**9** Most of the halogenated hydrocarbons find use in the chemical industry and medicine. The anesthetic halothane has the formula

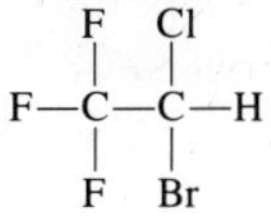

(III)

and the highly effective (and toxic) pesticide DDT has the structure

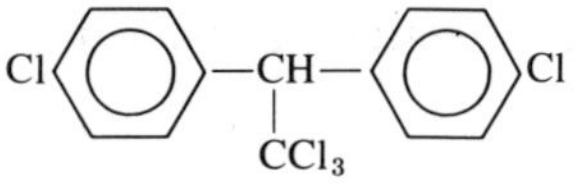

(IV)

## Section 2: ALCOHOLS

**1** *Alcohols* are hydrocarbons containing one or more *hydroxyl,* or —OH, groups. Thus, an alcohol contains C, H, and ________ atoms.

O (oxygen)

**2** We can give alcohols the generic formula or group designation ROH. Here the letter R stands for any hydrocarbon radical, the OH for the ________ group.

hydroxyl

**3** There are several systems used for naming alcohols. The simplest approach is to use the name of the radical plus the word alcohol. Thus, $CH_3OH$ becomes methyl alcohol; $C_3H_7OH$, ________ alcohol; and so on.

propyl

**4** In the IUPAC system, an alcohol is named from the alkyl radical and the suffix *-ol*. Thus, $CH_3OH$ is methanol, $C_2H_5OH$ is ethanol, and $C_3H_7OH$ is ________.

propanol

**5** Often it is necessary to designate the location of the OH group in the molecule. For example, propyl alcohol, $C_3H_7OH$, may have two different structures:

(V)

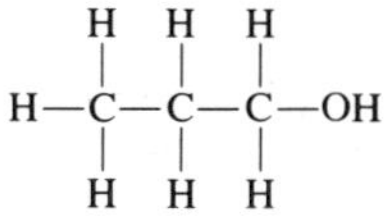

or

(VI)

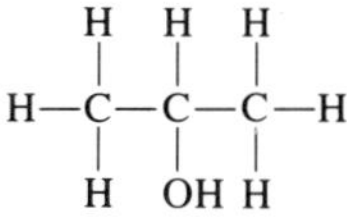

Structure V is normal or n-propyl alcohol, and structure VI is iso-
alcohol — propyl ______________.

**6** In the IUPAC system, normal propyl alcohol is *1-propanol* (OH on *first* carbon), and isopropyl alcohol is *2-propanol* (OH on *second* carbon). Because these alcohols have the same empirical formula, they
isomers — are ______________.

**7** Alcohols can be further characterized in terms of the structures of the hydrocarbon radical; that is, we have *primary, secondary,* and *tertiary* alcohols.

(VII)

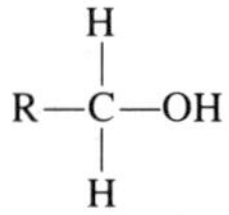

**Primary**

(VIII)

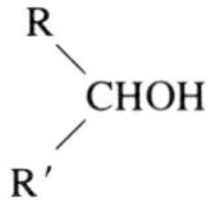

**Secondary**

(IX)

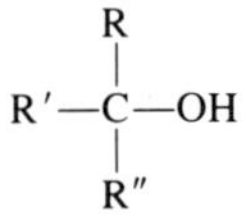

**Tertiary**

**8** Look at the *carbon* atom to which the OH is attached in each case above. For a primary alcohol, this carbon is attached to 1 organic radical only. For a secondary alcohol, it carries 2 radicals, which may be the same or different. In the tertiary alcohol, this carbon is
3 — attached to ________ organic radicals.

**9** Classify each of the following structures as a primary, secondary, or tertiary alcohol:

(X)

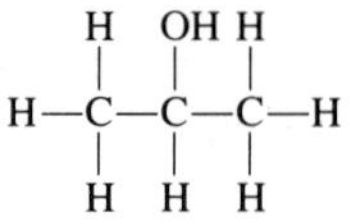

(XI)

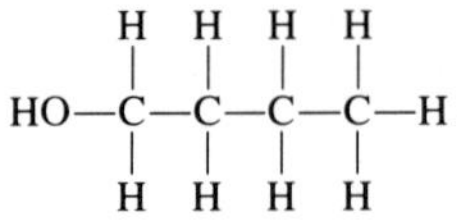

secondary; primary — ______________ ______________

(XII)

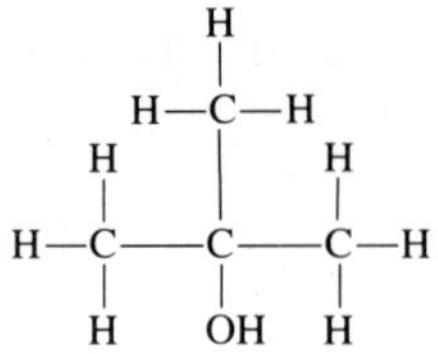

(XIII)

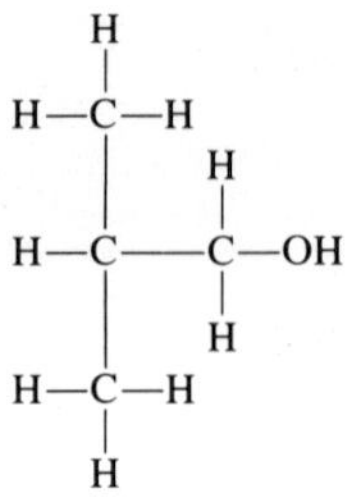

tertiary; primary — ______________ ______________

**10** Hydroxyl groups attached to aromatic rings produce a class of compounds called *phenols*. Phenol (carbolic acid), the simplest compound of the class, has the structure:

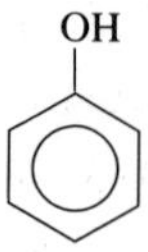

(XIV)

Phenol was the first antiseptic to be used in surgery, and its derivatives are widely used as disinfectants.

**11** Many alcohols are constituents of living cells; for example, the *triol* (3 OHs) glycerol

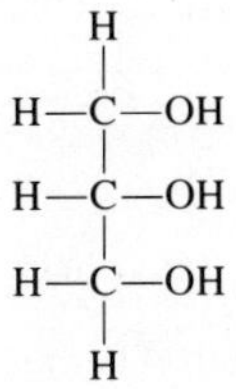

(XV)

is an important component of animal and plant fats and oils. Many alcohols are produced by fermentations catalyzed by microorganisms. Ethanol, or ____________ alcohol, is produced by the action of yeast enzymes on sugar.

ethyl

## Section 3: ORGANIC ACIDS

**1** Organic acids have the generic formula RCOOH. R stands for a kind of ________________ radical.

hydrocarbon

**2** The grouping —COOH is called a *carboxyl* group:

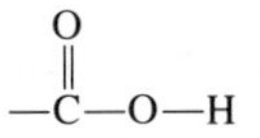

(XVI)

Thus RCOOH compounds are frequently called carboxylic ______.

acids

**3** Compounds containing the carboxyl group are acids because, like all acids, they undergo dissociation to yield free ________________ in water.

protons or hydrogen ions

(XI-3)

**4** Because this is a reversible process, we can predict that carboxylic acids will be *strong/weak* acids.

weak

**5** In everyday usage, most acids are known by common names. For example, we call $CH_3COOH$ acetic acid.

**6** Carboxylic acids react with bases to form salts and water; thus

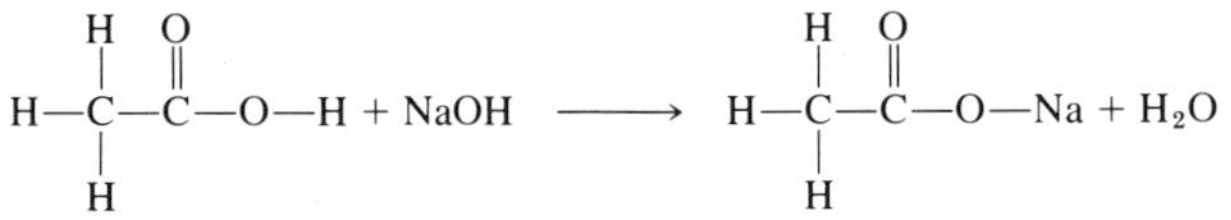

(XI-4)

**Acetic acid** **Sodium acetate**

**7** These neutralization reactions *are/are not* reversible.

are not

**8** The salts of carboxylic acids are salts of weak acids. If the base, as above, is a strong base, solutions of the salt will be *acidic/basic* in pH. If a weak base reacts with a carboxylic acid, the resulting salt will form a ________________ solution.

basic

neutral

## Section 4: ESTERS

**1** An *ester* is the reaction product of an acid and an alcohol. For example, acetic acid reacts with methyl alcohol as follows:

(XI-5)

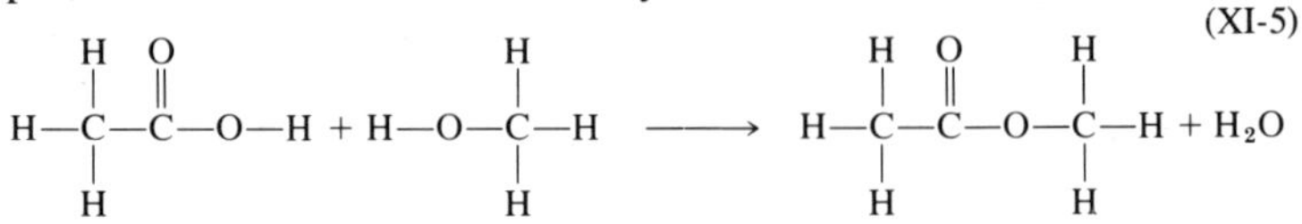

**Acetic acid** **Methyl alcohol** **Methyl acetate**

esters **2** Methyl acetate is one of a group of compounds called ____________, which have the generic formula RCOOR′, or

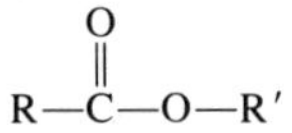

(XVII)

**3** The process shown in equation XI-5 is called *esterification*. It appears analogous to a reaction called

neutralization ____________, which we studied in inorganic chemistry.

**4** Esters are found in many living tissues. Many flavors and essences are esters; so also are fats and oils. All esters, however, share the

acid characteristic of being formed from a(n) ____________ and a(n)

alcohol ____________. All esters derived from carboxylic acids have

RCOOR′ the generic formula ____________.

## Section 5: ALDEHYDES

**1** Aldehydes have the generic formula RCHO:

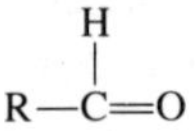

(XVIII)

They contain the carbonyl group —C=O, as do carboxylic acids and the esters of carboxylic acids. In aldehydes, only 1 hydrocarbon

carbonyl radical is attached to the ____________ group.

**2** Aldehydes are prepared by the *partial* oxidation of a *primary* alcohol; for example:

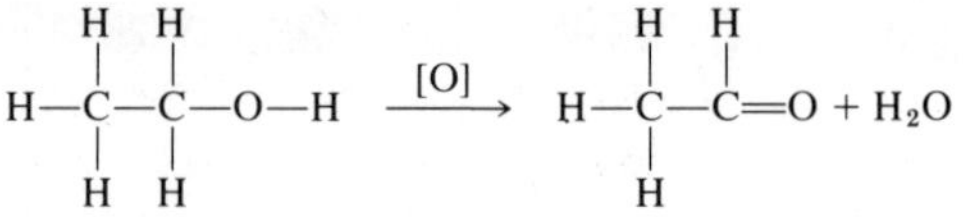

(XI-6)

**Ethyl alcohol** **Acetaldehyde**

hydrogen In other words, to form an aldehyde from an alcohol, 2 ____________ atoms must be removed.

**3** The oxidation above results in the formation of —C=O, the

carbonyl ____________ grouping.

**4** Further oxidation of aldehydes yields acids; for example:

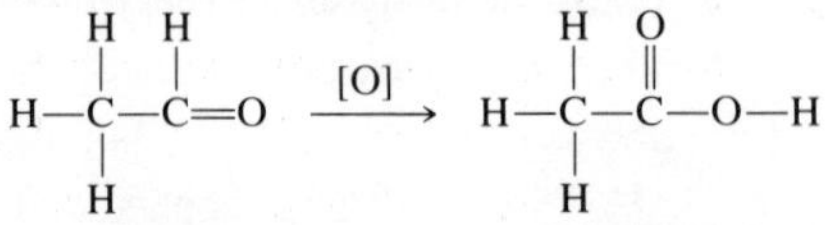

(XI-7)

**Acetaldehyde** **Acetic acid**

carboxylic The carbonyl group is here converted to a ________________ group.

**5** Aldehydes are commonly named after the acids they yield on oxidation. Thus, acetic acid is derived from acetaldehyde, and formic
formaldehyde acid from ________________. In the IUPAC system they carry the suffix *-al;* for example, acetaldehyde is *ethanal.*

**6** The smaller aldehyde compounds are very reactive molecules. Larger ones, including many sugars, are important to living organisms.

## Section 6: KETONES

**1** *Ketones,* with the generic formula RCOR′ and structural formula

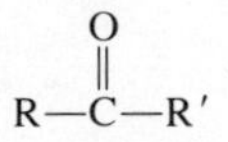

(XIX)

are closely related to aldehydes. They contain a carbonyl group
radicals linked to 2 hydrocarbon ________________.

**2** Ketones are prepared by the oxidation of *secondary* alcohols; for example:

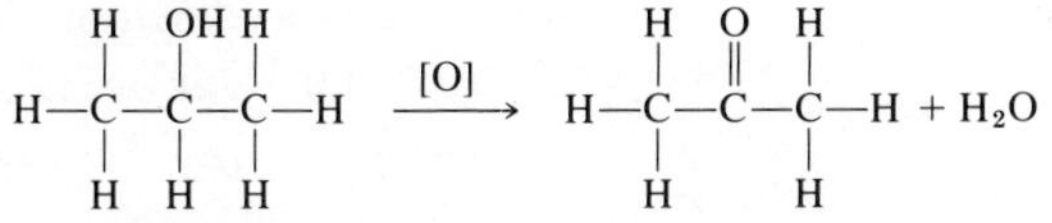

(XI-8)

**Isopropyl alcohol** **Dimethyl ketone**

hydrogen In this reaction, 2 ________________ atoms are removed to create
carbonyl a ________________ group.

**3** Simple ketones are named from the 2 radicals attached to the carbonyl group plus the word ketone; for example, dimethyl ketone, commonly known as acetone. In the IUPAC system ketones carry the suffix *-one;* for example dimethyl ketone is *propanone.*

## Section 7: ETHERS

**1** An *ether* is an organic oxide with the generic formula ROR′.

**2** If the Rs are the same, the compound is a simple *ether;* if they are different, it is a *mixed* ether. Therefore, dimethyl ether

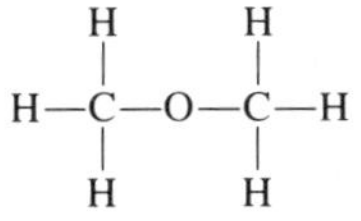

(XX)

simple is a ____________ ether.

**3** Methyl ethyl ether

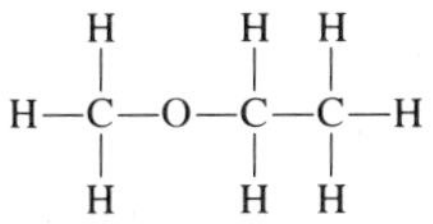

(XXI)

mixed must be a ____________ ether.

4 Diethyl ether, the well-known anesthetic ether, has the structure:

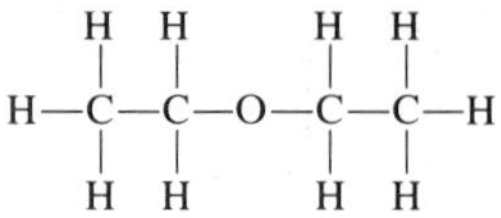

(XXII)

5 To summarize, hydrocarbons with *oxygen-containing substitutions* fall into six groups. Give the name of the class of compound for the following generic formulas:

| | | |
|---|---|---|
| ketone | RCOR′ | ________ |
| alcohol | ROH | ________ |
| acid | RCOOH | ________ |
| ester | RCOOR′ | ________ |
| ether | ROR′ | ________ |
| aldehyde | RCHO | ________ |

## Section 8: AMINES AND IMINES

1 Amines and imines are nitrogen-substituted hydrocarbons; that is, in addition to hydrogen and carbon, they contain ________.

nitrogen

2 Amines may be considered organic derivatives of ammonia, $NH_3$. One, 2, or all 3 of the hydrogens may be replaced by organic radicals. Where 1 hydrogen is replaced, we have a *primary* amine, $RNH_2$; 2 hydrogens, a secondary amine, $R_2NH$; 3 hydrogens, a ________ amine, $R_3N$.

tertiary

3 Therefore, we can class the structures below as:

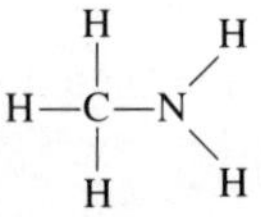

(XXIII)

primary ________ **amine**

(XXIV)

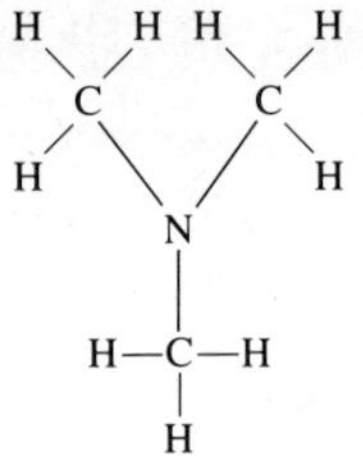

tertiary ________ **amine**

(XXV)

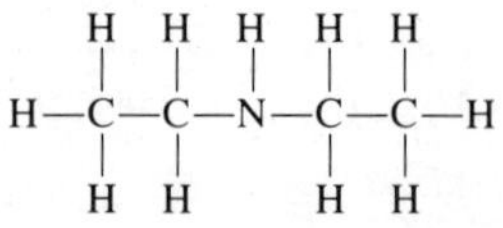

secondary ________ **amine**

4 Secondary amines are also known as *imines*. Imines, therefore, have a generic formula of ________.

$R_2NH$

**5** Nitrogen may also have a valence of 5. Tertiary amines react with alkyl halides to form *quaternary ammonium salts,* for example, tetramethyl ammonium chloride:

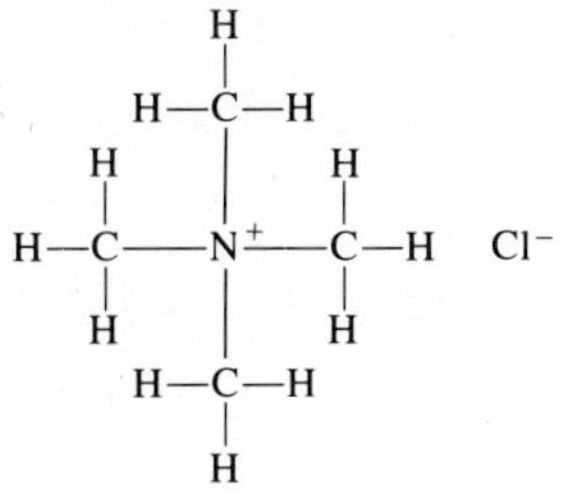

(XXVI)

Several compounds of this type, such as benzalkonium chloride (Zephiran), are widely used disinfectants.

## Section 9: CONDENSATION POLYMERS

**1** In Part X, Section 7, we became familiar with the concept of polymerization. Specifically, we studied polymers formed by simple *addition* of 1 hydrocarbon molecule to another; these are called

addition ________________ polymers.

**2** The addition type of polymer is formed from simple hydrocarbons. Substituted hydrocarbons form polymers by a more complex method called condensation polymerization.

**3** *Condensation polymers* are formed when monomers join together by the elimination of a *water* molecule. Thus, linking two monomers

hydrogen in a condensation polymer requires the removal of 2 ____________

oxygen atoms and 1 atom of ____________.

**4** Any reaction that joins 2 molecules together and eliminates a mole-

condensation cule of water is called a ________________ reaction.

**5** For example, certain amines and carboxylic acids can react by a condensation mechanism, as follows:

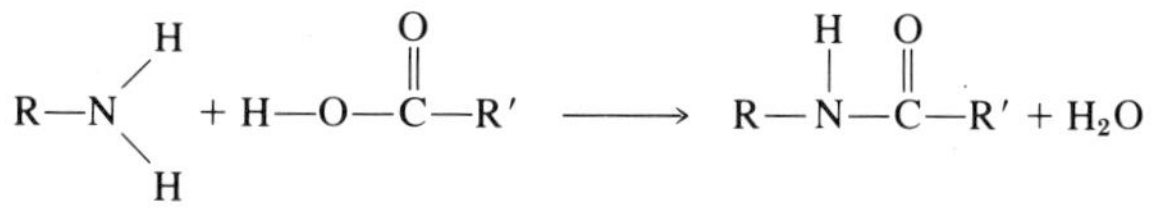

(XI-9)

**Amine** **Carboxylic acid**

condensation Because water is eliminated, this is a ________________ reaction.

**6** If the 2 participant molecules in a condensation reaction can react on both ends, a chain forms. Thus, in the product above the reaction of R with another carboxylic acid, and R′ with another amine, will

polymer produce a chain, or ________________.

**7** Whereas the addition polymers are important primarily as *man-made* industrial products, condensation polymers are vitally important both industrially and biologically. A living organism builds

condensation up its cellular materials by a controlled process of ____________ polymerization.

PART **XII**

# Biochemistry

## Section 1: CARBOHYDRATES

**1** *Biochemistry* is the chemistry of those compounds essential to life. A *biochemical* is a compound that is produced by and used by a living organism. All of these compounds contain carbon and are therefore ________ compounds.

organic

**2** Although there are more than a million different species of living things (plants, animals, bacteria, viruses, and so on), their cells all contain very similar groups of essential ________.

biochemicals

**3** Let us consider *carbohydrates* as our first group of biochemicals. These are compounds of carbon, hydrogen, and oxygen.

**4** Carbohydrates have an approximate generic formula of $(CH_2O)_n$. This shows that for each carbon atom there will usually be 2 ________ atoms and 1 atom of ________. This corresponds to 1 molecule of ________ for each carbon.

hydrogen; oxygen

water

**5** The simplest carbohydrates are *simple sugars*. These commonly have from 4 to 7 carbons. A 4-carbon sugar is a *tetrose*, a 5-carbon sugar a *pentose*, a 6-carbon sugar a ________, and a 7-carbon sugar a ________.

hexose

heptose

**6** Hexoses are by far the most common simple sugars. Let us look at the structures of two of the most common hexoses, or ____-carbon sugars.

6

(I)

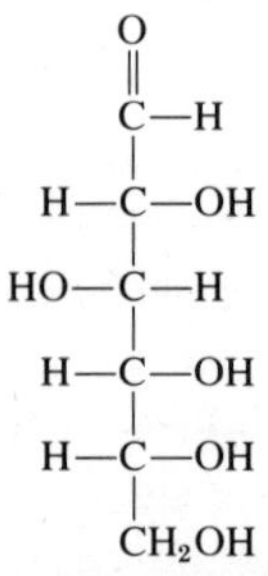

**Glucose** $(C_6H_{12}O_6)$

(II)

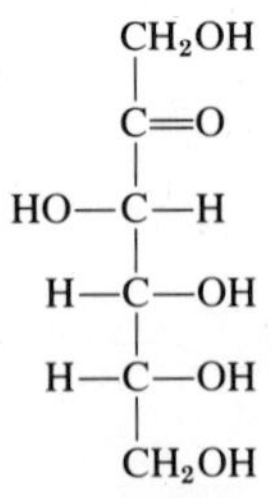

**Fructose** $(C_6H_{12}O_6)$

carbon

**7** You will notice that these hexoses have backbones of 6 ________ atoms.

alcohol

**8** Both compounds also have five —OH, or hydroxyl groups. Thus they can be considered to be a type of *acid/alcohol*.

ketone (carbonyl)

**9** Glucose, however, has an aldehyde group on one end, whereas fructose has a ________ group on the second carbon.

formula

**10** Glucose is an *aldehyde* sugar, and fructose is a *ketone* sugar. Note that they both have exactly the same empirical ________.

**11** The atoms of a glucose molecule may be bonded together in two alternate ways to satisfy valences:

(III)

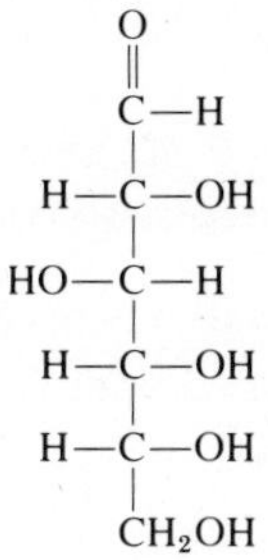

**Open form**

(IV)

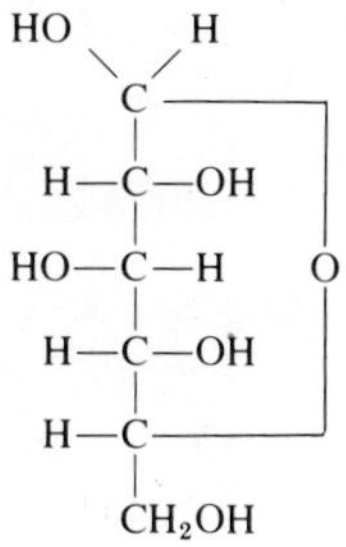

**Ring form**

(V)

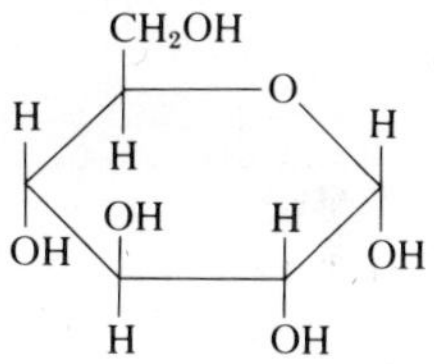

**Ring form**

ring

Structures III and IV represent two arrangements of atoms that exist in equilibrium; that is, 1 molecule of glucose may shift back and forth from the open form to the ________ form in solution.

carbon

**12** Structure V shows a commonly used structural drawing of the ring form. Notice that the ring consists of 6 atoms, 5 carbons and 1 oxygen. The ring does not include the sixth ________ atom.

ring

**13** Because ring forms are much more stable than open forms, practically all sugar molecules in solution will be found in the ________ form at any time.

polar

**14** Carbohydrates contain many polar covalent bonds. Many are therefore very soluble in *polar/nonpolar* solvents such as water and body fluids. Very large carbohydrates dissolve poorly in water, for mechanical reasons.

polysaccharides

**15** Simple sugars in the *ring form* undergo condensation reactions with each other to form larger carbohydrate molecules. Carbohydrates that are only 1 simple sugar per molecule are called monosaccharides; those made from 2 simple sugars are disaccharides. Carbohydrates containing many simple sugar units are called ________.

glucose

**16** There are only 3 disaccharides of nutritional significance. These are *maltose*, or malt sugar, with 2 glucose units; *lactose*, or milk sugar, containing a glucose and a galactose unit; and *sucrose*, cane or table sugar, with a glucose and a fructose unit. All 3 disaccharide molecules have in common the possession of at least 1 unit of ________.

**17** The equation for the formation of maltose requires 2 glucose molecules:

(XII-1)

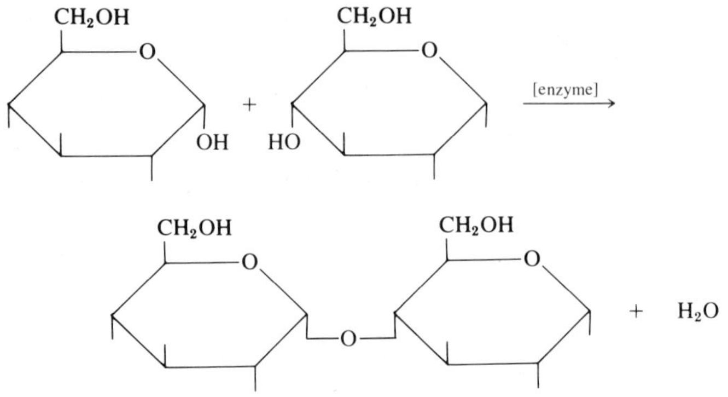


$$C_6H_{12}O_6 + C_6H_{12}O_6 \xrightarrow{\text{enzyme}} C_{12}H_{22}O_{11} + H_2O$$

**Glucose** **Glucose** **Maltose**

water

**18** Notice that in order to form a bond between the 2 glucose molecules, it is necessary to remove 1 ____________ molecule.

condensation

**19** Reaction XII-1 is therefore a ____________ reaction.

**20** In living cells, this reaction occurs only if catalyzed by an *enzyme*.

**21** Polysaccharides frequently contain several thousand sugar units. They are assembled by condensation and therefore are condensation

polymers

____________.

**22** The most important polysaccharides are starch, dextrin, glycogen, and cellulose. These are all polymers of glucose in which the glucose units are assembled in different ways. Their polymerization follows the equation

$$n\ C_6H_{12}O_6 \longrightarrow (C_6H_{10}O_5)_n + n\ H_2O \qquad \text{(XII-2)}$$

glucose

where n stands for the total number of molecules of ____________ that are required.

**23** Pectin, agar, carrageenin, and lignin are other biologically important sugar polymers, or ____________.

polysaccharides

**24** Sugars and starches are the primary source of food energy for humans and other animals. Cellulose (from wood) yields paper and fabrics.

**25** Give common examples of:

| | | |
|---|---|---|
| glucose, fructose, galactose | Hexose | ____________ |
| glucose, galactose | Aldehyde sugar | ____________ |
| fructose | Ketone sugar | ____________ |
| sucrose, lactose, maltose | Disaccharide | ____________ |
| starch, glycogen, cellulose | Polysaccharide | ____________ |

Other correct answers possible.

## Section 2: LIPIDS

**1** Lipids are a class of biochemicals of widespread distribution and importance. They have many different types of structures. The characteristic common to all lipids is solubility in nonpolar *organic solvents,* such as alcohols, acetone, or diethyl ether, rather than in polar solvents, such as water. Lipids, then, are generally not soluble

water — in ____________.

**2** For our purpose, the most important lipids are *fats* and *oils*. The major physical difference between fats and oils is that fats are solids at

liquids — room temperature, whereas oils are ________________.

**3** Chemically, both fats and oils are *esters* of glycerol and carboxylic acids. You will recall that the generic formula for an ester is

RCOOR′ — ________________.

**4** Because glycerol has 3 alcohol groups, it can esterify with 3

carboxylic acid — ________________ ____________ molecules. The resulting ester molecule is called a *triacylglycerol* (formerly called triglyceride).

**5** Carboxylic acids found in fats and oils are long-chain acids of more than 4 carbons. Most frequently, acids of from 12 to 20 carbon atom chain lengths are found. These are called *fatty acids*. A fatty acid is

4 — simply a carboxylic acid with more than ______ carbon atoms.

**6** Fatty acids contain many carbon and hydrogen atoms, plus 2 oxygen atoms. If all carbon atoms are linked by *single bonds,* the compound, like an alkane, is said to be *saturated*. You will recall that saturation means that the carbons are holding the maximum number of

hydrogen — ________________ atoms.

**7** A general formula for saturated fatty acids is $C_nH_{2n}O_2$. For example,

$C_{12}H_{24}O_2$ — 12-carbon fatty acid, *lauric acid,* is ____________, or, in the usual form, $C_{11}H_{23}COOH$.

**8** Structurally, we render lauric acid as

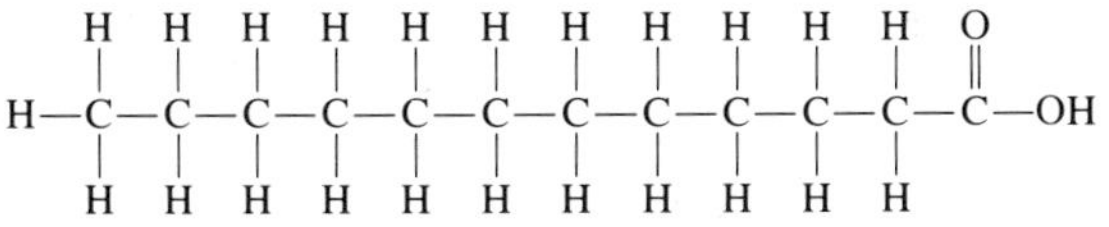

(VI)

or

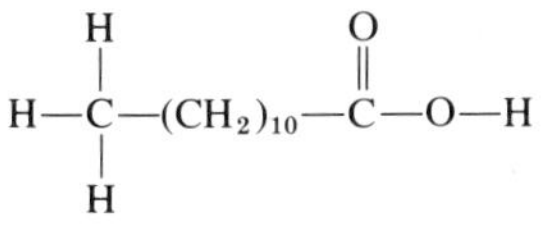

(VII)

Structure VII is a timesaving shortcut for structure VI.

single — **9** Note that in saturated fatty acids all carbons are joined by ________ bonds.

**10** *Unsaturated* fatty acids contain 1 or more carbon-to-carbon *double*

hydrogen — *bonds* and therefore less than the maximum number of __________ atoms.

**11** An example of an unsaturated fatty acid is *oleic acid*. Like stearic acid, it has 18 carbons, but 1 double bond:

(VIII)

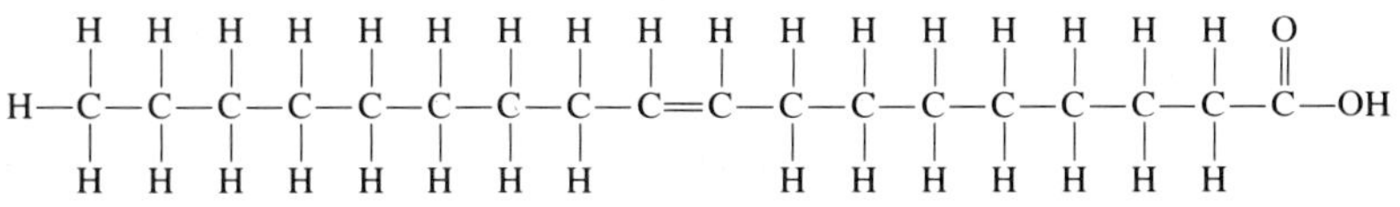

solids

**12** Fats and oils contain 3 fatty acid *residues* esterfied to a glycerol molecule. All three may be saturated or unsaturated, or they may be a mixture. We can generalize by saying that fats, which are ________ at room temperature, usually are esters of *saturated* fatty acids.

fatty acids

**13** Oils, which are liquids at room temperature, usually are esters of unsaturated ________ ________.

glycerol

**14** Most naturally occurring fats and oils, such as butter, beef suet, and peanut oil, are mixtures of different triacylglycerols, each of which in turn may contain 1, 2, or 3 fatty acids. To clarify this point, recall that a fat or oil is an ester of *1* molecule of ________ and *3* molecules of fatty acid (the three molecules may be the same or different).

**15** A typical triacylglycerol found in animal tissues is glyceryl tristearate *(tristearin),* an ester of glycerol and stearic acid. Its structure is:

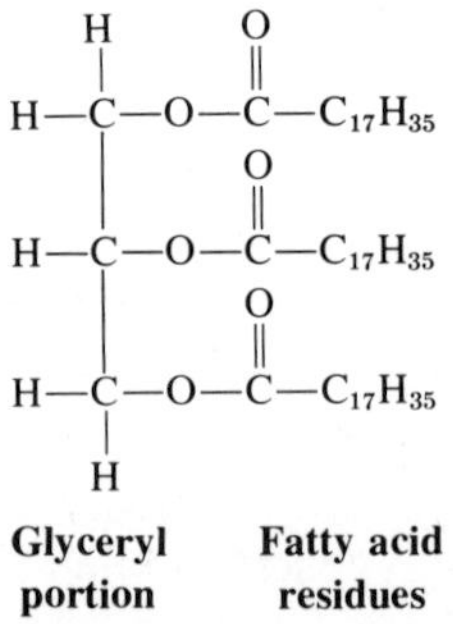


(IX)

saturated

**16** Beef fat contains glyceryl tristearate and other triacylglycerols. These are esters of *saturated/unsaturated* fatty acids.

**17** The human body can manufacture all necessary saturated fatty acids; however, it cannot manufacture certain key unsaturated fatty acids.

diet

**18** Certain unsaturated fatty acids must be supplied as part of a balanced ________.

energy

**19** Fats and oils are used by living organisms primarily as sources of reserve food energy. These compounds contain considerable potential chemical ________.

cation

**20** Closely related to fats and oils are *soaps.* A soap is the sodium or potassium salt of a fatty acid. Thus, 1 soap molecule contains a fatty acid *anion* and a monovalent metallic ________.

glycerol

**21** There are several classes of complex lipids related to fats and oils. Typically, they are glycerol esters that contain 1 or 2 fatty acid residues and 1 or 2 other groups esterified to the glycerol backbone. For example, *phospholipids* such as lecithin and cephalin are composed mainly of fatty acids, phosphoric acid, and ________.

permeability

**22** Phospholipids are major components of cell membranes. Their characteristics contribute to the membrane's semi________.

organic (nonpolar)

water

**23** There are also lipids that function as vitamins and hormones. Their structures are quite different from those discussed above, but like all lipids they are soluble in ________ solvents and insoluble in ________.

## Section 3: PROTEINS

**1** Proteins are a critically important class of biochemicals and are the predominant compounds in practically all living tissue. Chemically, they contain carbon, hydrogen, oxygen, nitrogen, and sulfur. The most striking feature of proteins is their ability to function as biological catalysts, or ________.

enzymes

**2** Proteins are condensation polymers of *amino acids*. There are *20* naturally occurring amino acids found in proteins. Their generic formula is:

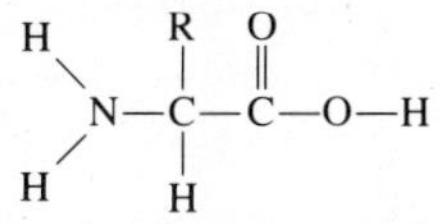

(X)

Note that this structure contains the —$NH_2$, or amino group, characteristic of primary ________.

amines

**3** The generic formula also contains the —COOH found in all carboxylic ________.

acids

**4** In the above generic formula, R represents any one of 20 different organic radicals. Thus, 1 amino acid differs from another chiefly in terms of which organic ________ is found at R.

radical

**5** Different R groups can give amino acids different properties. For example, if R is a *nonpolar* hydrocarbon, the amino acid will be poorly soluble in water; on the other hand, if it contains a hydroxyl group, the amino acid will be very soluble in water. The properties of a protein are determined by the properties of its constituent ________ ________.

amino acids

**6** Amino acids are *amphoteric,* which means that they can function as both acids and ________.

bases

**7** The carboxyl group functions as an acid; that is, it can donate a ________ and become $COO^-$.

proton

**8** The amino group functions as a base; it can accept a ________ and become $NH_3^+$.

proton

**9** Thus, an amino acid can have both positive and negative ________ charges when dissociated.

electrical

**10** Several amino acids have R groups that contain carboxylic acid groups. These are called *acidic amino acids* and may have an excess of *positive/negative* charges.

negative

**11** If the R side chain contains amino groups, there may be an excess of positive charges in the amino acid. This will make the amino acid *basic/acidic*.

basic

**12** The dissociation of both amino and carboxyl groups is affected by pH. For example, carboxyl groups are more likely to dissociate at a high pH, whereas amino groups will be in the charged form (—$NH_3^+$) at a ________ pH.

low

**13** Each protein has a different combination of amino acids. If there are more acidic amino acids than basic ones, the protein itself will be ________, and vice versa.

acidic

**14** Each protein has some pH at which the numbers of negatively charged —$COO^-$ ions and positively charged —$NH_3^+$ ions are equal. This pH is called the *isoelectric point* for the protein. It is the point at which the protein is electrically ________________.

neutral

**15** The amphoteric quality of proteins is used in separating protein mixtures. Blood serum and other body fluids are colloidal mixtures of acidic, basic, and ________________ proteins.

neutral

**16** If we place a blood serum sample in an electrical field, the negatively charged, or ________________, proteins will move toward the *anode*, and positively charged, or basic, proteins will move toward the ________________.

acidic

cathode

**17** Neutral proteins are largely unaffected by a(n) ________________ ____________.

electrical field

**18** The separation of a mixture of proteins by means of an electrical field is called *electrophoresis*. The procedure depends on the fact that some proteins carry an electrical ________________.

charge

**19** Proteins are polymers of ____________ ____________.

amino acids

**20** Amino acids are linked together to form proteins by means of condensation polymerization, a reaction in which ____________ is removed.

water

**21** The basic reaction in the condensation of amino acids is:

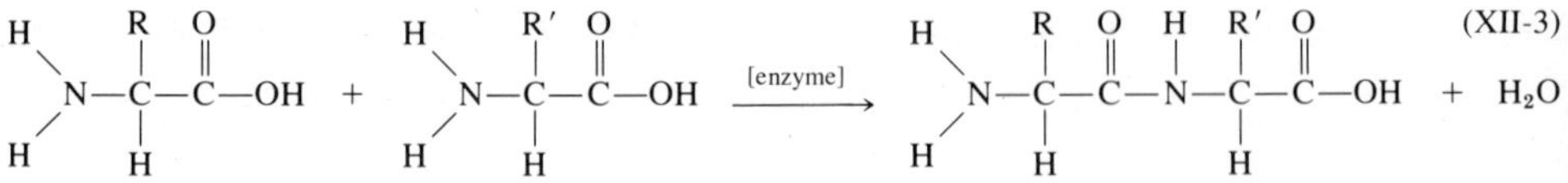

(XII-3)

Note the bond formed between the *carboxyl carbon* of 1 amino acid and the *amino* ________________ of the other.

nitrogen

**22** The resulting compound (XII-3) is a peptide, a small group of amino acids. The bond formed, although it is the familiar covalent bond, is given the special name of *peptide bond*. A peptide bond is a covalent bond linking 2 ____________ ____________.

amino acids

**23** Peptide bond formation in nature requires an enzyme catalyst and is under *genetic control*. When several amino acids have been linked together we can call the product a polypeptide. Complex, functional polypeptides are ________________.

proteins

**24** Genetic information tells the cell *how many* amino acids and in *what order* they are to be used to make a certain protein. Thus, each protein has its own unique *amino acid sequence* and length, which is under ________________ control.

genetic

**25** Albumin, the protein of egg white, has a different ____________ ____________ ________________ than keratin, the protein of hair.

amino acid sequence

**26** The amino acid sequence of a protein is like a sentence: each sentence has its own ________________ of words.

sequence

**27** The amino acid sequence of a protein is called its *primary structure*.

**28** Protein chains may interact to form secondary structures—either flat sheets or springlike coils called alpha helixes. These structures are stabilized by the formation of weak bonds called *hydrogen bonds*.

covalent

**29** You will remember that atoms of unequal electronegativity form polar ________________ bonded molecules.

negative

**30** *Polar* molecules have one end that is electrically positive and one that is electrically ________________.

negatively

**31** A *hydrogen bond* is a weak attraction formed when the positively charged end of 1 polar molecule is attracted to the ____________ charged end of another polar molecule.

secondary

**32** Hydrogen bonds are important in maintaining the *primary/secondary* structure of proteins.

proteins

**33** In addition to primary and secondary structures, some proteins have a *tertiary* structure, or third level of organization. These are alpha helix compounds that undergo folding to form a compact, roughly spherical package called a *globular protein*. Most enzymes are globular ________________.

amino acids

**34** Let us summarize some important points. Proteins are condensation polymers of ____________ ____________.

basic

**35** Amino acids may be soluble or insoluble, acidic or ____________.

neutral

**36** Depending on which amino acids are present, a protein itself may be acidic, basic, or ________________ in pH.

amino acid sequence

**37** A protein's primary structure, which is under genetic control, is its ____________ ____________ ________________.

secondary

**38** Interactions between amino acids within the protein may result in hydrogen bonds and thereby stabilize a ________________ level of protein structure.

tertiary

**39** A globular protein is one in which folding of the helix has produced a ________________ structure.

genetic

**40** The unique pattern and structure of each different protein is dictated by ________________ information.

## Section 4: NUCLEIC ACIDS

monomers

**1** The fourth major class of biochemicals is the *nucleic acids*. Like polysaccharides and proteins, nucleic acids are polymers, that is, molecules made up of repeating units, or ________________.

**2** The monomer of a nucleic acid is called a *nucleotide*. Each nucleotide in turn is made up of three subunits: a *nitrogenous base*, a *sugar*, and a *phosphate* group.

**3** Structure XI shows a representative nucleotide in outline, and structure XII shows atomic detail:

(XI)

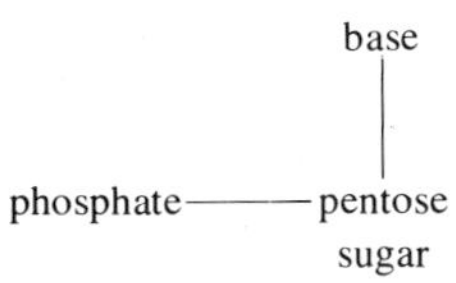


(XII)

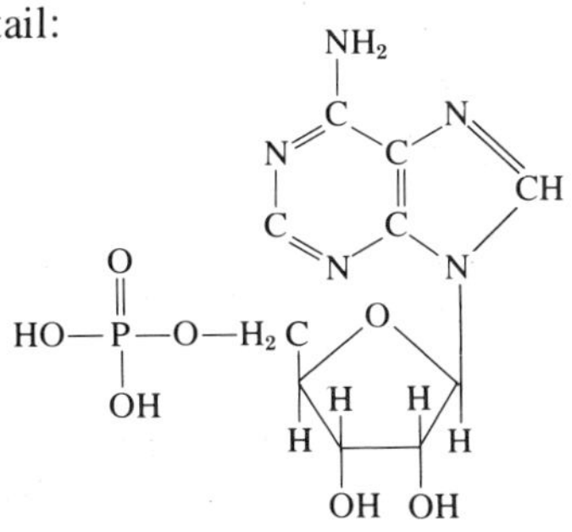


**Adenosine monophosphate**

**Table 7.** Nucleic acids and their nucleotides

| *Nucleic acid* | *Base* | *Sugar* | *Phosphate* | *Nucleotide* | *Monophosphate (abbreviation)* | *Triphosphate (abbreviation)* |
|---|---|---|---|---|---|---|
| DNA | Adenine | Deoxyribose | Present | Deoxyadenosine-5′-phosphate | dAMP | dATP |
| | Guanine | Deoxyribose | Present | Deoxyguanosine-5′-phosphate | dGMP | dGTP |
| | Cytosine | Deoxyribose | Present | Deoxycytidine-5′-phosphate | dCMP | dCTP |
| | Thymine | Deoxyribose | Present | Deoxythymidine-5′-phosphate | dTMP | dTTP |
| RNA | Adenine | Ribose | Present | Adenosine-5′-phosphate | AMP | ATP |
| | Guanine | Ribose | Present | Guanosine-5′-phosphate | GMP | GTP |
| | Cytosine | Ribose | Present | Cytidine-5′-phosphate | CMP | CTP |
| | Uracil | Ribose | Present | Uridine-5′-phosphate | UMP | UTP |

The nucleotide shown is adenosine monophosphate. The adenine base is *heterocyclic/carbocyclic*.

heterocyclic

**4** The above base is covalently linked to a pentose, a sugar with ______ carbons.

5

**5** The two known classes of nucleic acids differ in regard to their pentose sugar (Table 7). Thus, *deoxyribonucleic acid* (DNA) contains the sugar deoxyribose, whereas *ribonucleic acid* (RNA) contains the sugar ______________.

ribose

**6** The same bases are found in both DNA and RNA, with the exception that only DNA has ______________ and only RNA has ______________.

thymine
uracil

**7** Nucleotides derived from nucleic acids by chemical degradation always have *1 phosphate* attached to the *fifth carbon* of the pentose sugar. They are therefore *monophosphates*. This explains the origin of the abbreviation AMP for adenosine ______________.

monophosphate

**8** Nucleotides, as the cell uses them for the synthesis of nucleic acids and for other purposes, always have a string of *3* phosphates attached to the sugar. These are *triphosphates*. Thus, dCTP stands for ______________ ______________.

deoxycytidine triphosphate

**9** Any nucleotide with the prefix deoxy- is a unit of *RNA/DNA*.

DNA

**10** If a nucleotide has no prefix it is a unit of ______________.

RNA

**11** The nucleotide monomers are joined by a series of enzymatic condensation reactions to form long nucleic acid ______________.

polymers

**12** Nucleotides condense as follows:

$$2\left[\begin{array}{r}\text{base}\\ |\\ \text{triphosphate—sugar}\end{array}\right] \xrightarrow{\text{[enzyme]}} \begin{array}{c}\text{base}\qquad\qquad\qquad\qquad\quad\text{base}\\ |\qquad\qquad\qquad\qquad\qquad\quad |\\ \text{triphosphate—sugar—phosphate—sugar}\end{array} + PP_i + H_2O \qquad \text{(XII-4)}$$

Note that water and 2 phosphates are released in this reaction.

**13** The structure of RNA is determined almost entirely by the polymerization process. RNAs are long single strands of polymerized ________________.

nucleotides

**14** The important feature that makes one RNA different from another is its *base sequence*.

**15** There are four different nucleotides, which can be put together in any order. The order of the nucleotides then is indicated by the nitrogen base ________________.

sequence

**16** Structurally, DNA is considerably more complicated than RNA. Its complexities are of overwhelming importance because within its structure resides the master *genetic code*.

**17** Natural DNA is made up of two strands of polymerized nucleotides, rather than one strand as in ____________.

RNA

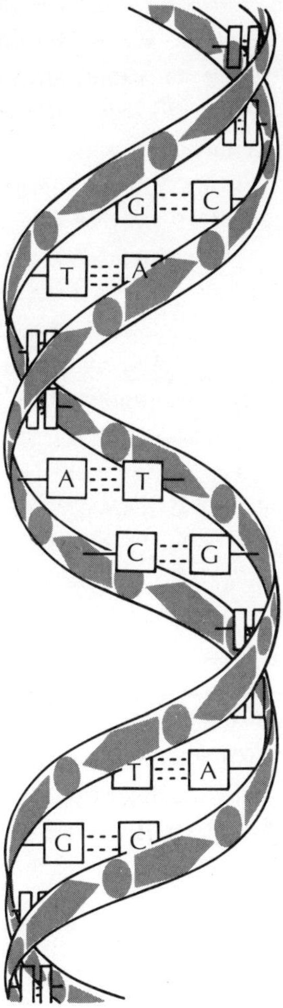


**Fig. 4.** The DNA double helix. Gray symbols of strands represent sugar and phosphate residues; letters and dotted lines represent bases and hydrogen bonds respectively. (From Cohen, A.: Handbook of cellular chemistry, St. Louis, 1975, The C. V. Mosby Co.)

dTMP

**18** Each strand of DNA contains a sequence of 4 deoxyribonucleotides: dAMP, dCMP, dGMP, and ________.

**19** In turn, the two strands of DNA are held together by the establishment of *hydrogen bonds* between the bases of the two strands.

cytosine

**20** It has been shown that adenine forms hydrogen bonds *only* with thymine, and that guanine bonds *only* with ________.

strands

**21** *Specific base pairing* is involved in bonding together the two ________ of DNA.

specific base

**22** Fig. 4 shows how these interactions work. Note that the sugars and phosphates form the backbone for each strand. The ladderlike cross-connections are formed by ________ ________ pairing.

two strands

**23** The nucleotides making up the DNA molecule are bonded together in such a way that the ________ ________ wrap around each other in the form of a *double helix*.

complementary

**24** You will notice in Fig. 4 that one strand matches the other; that is, the bases of one strand are exactly those that will base-pair with those of the other. Normally there are no mismatches, and the two strands are said to be *complementary/identical*.

sequence

**25** The DNA sequence of bases spells out the genetic code, which in turn is enzymatically *transcribed* into the base ________ of *messenger* RNA (mRNA).

amino acids

**26** The base sequence of mRNA is enzymatically *translated* into protein. More particularly, mRNA serves as a pattern, or *template,* for the assembling of ________ ________ into protein.

**27** A special type of RNA, called *transfer* RNA (tRNA), has the job of taking the various amino acids to the mRNA template.

template

**28** In genetic *translation,* tRNAs, each coupled with a particular amino acid, are aligned along the mRNA ________, according to their matching bases.

messenger
transfer
protein

**29** A gene may be viewed as a particular *segment* of a DNA strand whose base sequence becomes transcribed into a complementary base sequence of ________ RNA, and thereby (with the assistance of ________ RNA) is translated into a particular ________.

genetic

**30** When a cell divides, the two strands of DNA unwind and serve as templates for the synthesis of complementary strands in the new (daughter) cells identical to their partner strands in the parent cell. Thus, the daughter cells and the parent cell are alike because all three possess the same ________ code.

# PART XIII

# Metabolism

## Section 1: METABOLIC PROCESSES

1 *Metabolism* is the total of the biochemical processes that take place within the living organism. As a simplification, we can divide the hundreds of reactions of metabolism into two areas, catabolism and anabolism. *Catabolism* is the *breaking* of the chemical bonds in large, complex molecules to form *smaller, simpler* molecules; *anabolism* is the *synthesis (building up)* of larger molecules from simpler molecules.

anabolism

2 Some of the products of catabolism serve as raw materials for the process of ________________.

exergonic

3 In general, most catabolic reactions yield energy; that is, they are *exergonic/endergonic*.

catabolic

4 Anabolic reactions are those in which the cell or organism assembles small molecules, or "building blocks," into macromolecules. For example, one important anabolic process is the synthesis of proteins from the amino acids furnished by ________________ reactions.

endergonic

5 Anabolic reactions generally require an *energy input;* that is, they are ________________.

chemical

6 Being endergonic, anabolic processes serve to store energy. For example, the synthesis of fat results in the accumulation of potential ________________ energy.

catabolism

7 Balanced metabolism implies a balance between anabolism and ________________.

8 Because catabolism and anabolism involve hundreds of reactions, chemical coordination and *balance* are critical. The maintenance of this balance is called *homeostasis*.

homeostasis

9 When the *rates* of all metabolic processes in an organism are correctly adjusted, we can say that a state of ________________ exists.

## Section 2: ENZYMES

1 *Enzymes* are biological *catalysts*. They regulate the rates of all metabolic processes.

enzymes

2 Homeostasis is maintained through the rate-controlling functions of many hundreds of ________________.

**3** Review Sections III and IV of Part III before proceeding further.

**4** Each chemical reaction proceeds at a characteristic rate. In the laboratory this rate can be influenced by catalysts, or by changes in
concentrations
pressure, temperature, or ________________ of the reactants.

**5** In a living organism, we cannot vary temperature or pressure without doing damage. There is also very little allowable fluctuation in the
reactants
concentrations of the ________________.

**6** Therefore, the major way in which a metabolic reaction rate may be
enzyme
altered is through the participation of a(n) ________________.

activation
**7** Enzymes, like other catalysts, alter the ________________ energy of a reaction.

**8** Homeostasis requires the fine adjustment of each of hundreds of reaction rates. To achieve this, the organism must be able to control each reaction individually. This is why in living cells there is one
enzyme
specific catalyst, or ________________, for each reaction.

**9** We use the phrase *enzyme specificity* to describe the fact that typically
reaction
an enzyme catalyzes a specific ________________.

**10** Although it may be difficult to visualize, a cell that carries out 2,000
2,000
chemical processes may be expected to contain ________________ different enzymes.

**11** All known enzymes are *proteins*. Each different enzyme is a chemically distinct protein that has its own unique sequence of
amino acids
____________ ____________. Most functional enzymes have a tertiary structure of folded alpha helixes and are, therefore, of the
globular
________________ protein type.

**12** We can view an enzyme, then, as a three-dimensional structure with an uneven surface covered with folds and ridges. Each different enzyme will have a different pattern of folds and ridges, or *surface configuration*.

**13** Enzymes are specific because they recognize and combine with the reactants for their own particular reaction. A reactant in an enzyme-catalyzed reaction is called a *substrate*. Thus, the biochemical
substrate
an enzyme combines with in a reaction is its ________________.

**14** We can explain enzyme specificity by pointing out that each enzyme has a special shape and surface configuration; that is, it fits physically and chemically with a particular substrate that has its own specific molecular structure. An enzyme and its substrate form an *enzyme-substrate complex*.

**15** We can describe an enzyme-catalyzed reaction as follows:

$$E + S \rightleftharpoons ES \rightleftharpoons E + P \qquad \text{(XIII-1)}$$

That is,

$$\text{enzyme + substrate} \rightleftharpoons \text{complex} \rightleftharpoons \text{enzyme + product}$$

Note that the enzyme molecule emerges *unchanged*. This is charac-
catalyst
teristic of a true ________________.

**16** Not only do E + S associate to yield the ES complex, but ES can
dissociate
________________ to yield E + S.

**17** Two reversible steps are involved in reaction XIII-1; therefore, the process has a *complex* equilibrium. An enzyme-catalyzed reaction adjusts its net rate to conditions within the organism. If product (P) is in very *low* concentration, the rate of product formation will *increase*. If there is an *excess* of product, more than the cell can use, the rate of product formation will ________.

decrease

**18** Thus, metabolic processes adjust themselves to the needs of the organism. Each reaction has its own enzyme, and each is *self-adjusting*. This is the essential principle of balanced metabolism, or ________.

homeostasis

**19** Many enzymes in higher organisms are manufactured in an inactive form called a *proenzyme,* or *zymogen*. For example, pepsinogen is the ________ form of the digestive enzyme pepsin.

inactive

**20** Proenzymes are activated by other enzymes called *kinases*. Thus, when active trypsin is formed from inactive trypsinogin, a ________ enzyme is acting.

kinase

**21** Some enzymes require nonprotein helper materials for activity. These helpers may be free ions, such as magnesium, potassium, or phosphate, which are called *inorganic/organic cofactors;* they may be small organic molecules called *coenzymes;* or they may be organic groups or metal atoms firmly incorporated into the three-dimensional structure of the protein *(prosthetic groups)*.

inorganic

**22** An ever-increasing number of enzymes have been found to require loosely-bound metal ions, or inorganic ________. In addition, many so-called metalloenzymes contain metal atoms as integral parts of their structure, that is, as ________ ________.

cofactors

prosthetic groups

**23** Nutritionists accordingly emphasize that *trace metals* or minerals, including iron, zinc, copper, and cobalt, are critical components of a balanced ________.

diet

**24** Loosely attached, small organic cofactors are called *coenzymes*. They are often, but not always, of nucleotide structure, and most, but not all, are derivatives of one or another of the important dietary essentials called *vitamins*.

**25** If the organic helper is *loosely* attached to the protein portion of the enzyme, it is called a ________; if it is a *firmly* attached component of the molecule, it is a ________ ________.

coenzyme

prosthetic group

**26** The term *apoenzyme* is sometimes applied to the main portion of the enzyme molecule, that is, the ________ portion.

protein

**27** A so-called *simple* enzyme is an enzyme protein that has no prosthetic group, coenzyme, or inorganic ________. A conjugated enzyme, in contrast, contains one or more of these.

cofactor

**28** The complete conjugated enzyme is called a *holoenzyme*. A holoenzyme contains some kind of cofactor in combination with the protein, or ________ portion.

apoenzyme

**29** In modern terminology, enzyme names have the suffix *-ase*. The

**Table 8.** Common types of enzymes

| *Enzyme type* | *Function* |
|---|---|
| Oxidase | Addition of oxygen atoms |
| Hydrogenase | Addition of hydrogen atoms |
| Dehydrogenase | Removal of hydrogen atoms |
| Hydrolase | Hydrolysis (addition of water to split molecules) |
| Phosphorylase | Phosphorolysis (addition of phosphate groups to split molecules) |
| Phosphatase | Removal of phosphate groups |
| Carboxylase | Addition of carbon dioxide |
| Decarboxylase | Removal of carbon dioxide |
| Isomerase | Rearrangement of atoms within molecules |
| Transaminase | Transfer of amino groups |
| Polymerase | Polymerization |

substrate

first part of the name may indicate either the *type* of reaction catalyzed or the *substance* acted on, that is, the ________. Examples of groups of enzymes named according to the type of reaction catalyzed are given in Table 8.

## Section 3: VITAMINS

1 *Vitamins* are relatively complex organic substances that occur naturally in plant and animal tissues and are essential in small amounts for the control of metabolic processes.

coenzymes

2 Table 9 shows the relationships of certain vitamins to the ________ of various segments of metabolism.

vitamins

3 Although all living things require the same coenzymes to function, not all require the same vitamins. This seeming paradox relates to the abilities of various organisms to manufacture few, many, most, or all of their own coenzymes from simpler materials. Essential dietary nutrients, therefore, may not include many ________.

diet

4 In human nutrition, most vitamins must be supplied intact in the ________; certain vitamins, namely D and K, can be synthesized in the body.

coenzymes

5 A *vitamin deficiency* can easily lead to disease because of the resulting lack of certain key ________ essential to normal metabolism.

water

6 Vitamins are commonly classified as *water soluble* and *fat soluble*. Water-soluble vitamins include the B vitamins (Table 9) and vitamin C. They are found in moist foods such as fruits and vegetables, which contain much ________.

fat

7 The fat-soluble vitamins A, D, E, and K, are found in foods such as nuts, grains, and butter, which contain much ________.

coenzyme A

8 A *balanced diet,* yielding adequate amounts of all essential vitamins, produces a balanced metabolism. For example, the B vitamin pantothenic acid serves as the precursor of ________.

**Table 9.** Water-soluble B vitamins

| *Vitamin* | *Coenzyme* | *Metabolic function* |
|---|---|---|
| Thiamine ($B_1$) | Thiamine pyrophosphate | Aerobic metabolism of carbohydrates |
| Riboflavin ($B_2$) | Flavin adenine dinucleotide | Oxidation-reduction, electron transport |
| Pyridoxine ($B_6$) | Pyridoxal phosphate | Amino acid metabolism |
| Niacin | Nicotinamide adenine dinucleotide | Oxidation-reduction, electron transport |
| Pantothenic acid | Coenzyme A | Metabolism of fats and carbohydrates |
| Biotin | Biotin | Fixation of carbon dioxide |
| Folic acid | Tetrahydrofolate | Amino acid and nucleotide synthesis |
| Cyanocobalamin ($B_{12}$) | Cobamide coenzymes | Amino acid and nucleotide synthesis |

## Section 4: CARBOHYDRATE METABOLISM

polysaccharides

**1** Carbohydrate metabolism is the *digestive* (hydrolytic) breakdown of double sugars (disaccharides) and starches (________________) and the fates of the end products of these digestive reactions.

glucose

**2** In the process of digestion, starches (the most abundant dietary carbohydrates) are enzymatically degraded to the monomer ________________, according to the summarized equation:

$$(C_6H_{10}O_5)_n + n\ H_2O \xrightarrow{\text{enzyme}} n\ C_6H_{12}O_6 \qquad \text{(XIII-2)}$$

hexoses

**3** Also in the digestive tract, disaccharides are enzymatically degraded to 6-carbon monosaccharides, or ________________, according to the equation:

$$\underset{\textbf{Disaccharide}}{C_{12}H_{22}O_{11}} + H_2O \xrightarrow{\text{enzyme}} \underset{\textbf{Monosaccharide}}{2\ C_6H_{12}O_6} \qquad \text{(XIII-3)}$$

water

**4** The above reactions, like all digestive reactions, involve hydrolysis; for every disaccharide unit degraded, 1 molecule of ________ is consumed in the reaction.

2

**5** The digestion of the disaccharide maltose yields *glucose;* specifically, 1 molecule of maltose yields ________ molecules of glucose.

glucose

**6** One molecule of the disaccharide sucrose is digested to 1 molecule of fructose and 1 molecule of ________________.

glucose

**7** One molecule of the disaccharide lactose yields 1 molecule of galactose and 1 molecule of ________________.

galactose
glucose

**8** The end products of carbohydrate digestion, then, include the monosaccharides glucose, fructose, and ________________; of these, the major product is ________________.

hexose

**9** Furthermore, fructose and galactose are ultimately converted into glucose; therefore, the metabolism of carbohydrates is essentially the biochemical fate of the ________________ sugar, glucose.

**10** Glucose serves as the body's chief fuel. Its energy (trapped among its 6 carbons) is released through a complex series of reactions collectively referred to as *cellular respiration*. Overall, cellular respiration is an *endergonic/exergonic* process.

exergonic

**11** The *first stage* of cellular respiration, *glycolysis*, is a sequence of nine enzyme-catalyzed reactions. It commences with the conversion of glucose to *glucose phosphate* (glucose P), which is then degraded to 3-carbon pyruvic acid:

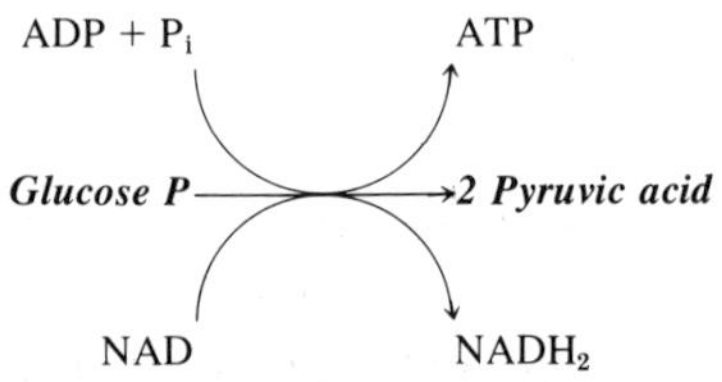


(XIII-4)

**12** Equation XIII-4 has been designed to show that three separate but intimately related processes are occurring. First, consider the conversion of glucose to ________________ acid.

pyruvic

**13** Pyruvic acid is a key metabolic intermediate, ($CH_3$—CO—COOH, or $C_3H_4O_3$). Because it contains 3 carbons, 2 molecules of pyruvic acid can be formed from glucose, which has ________ carbons.

6

**14** If we try to balance the equation

$$\begin{array}{ccc} \text{glucose} & \longrightarrow & \text{2 pyruvic acid} \\ C_6H_{12}O_6 & \longrightarrow & 2\ C_3H_4O_3 \end{array} \qquad \text{(XIII-5)}$$

we find 4 ________________ atoms left over.

hydrogen

**15** The fate of the hydrogen atoms can be explained when we realize that they have been transferred to *nicotinamide adenine dinucleotide* (NAD), a coenzyme derived from the vitamin ________________.

niacin

**16** NAD functions here and at many other points in metabolism as a hydrogen acceptor:

$$NAD + 2H \rightleftharpoons NADH_2 \qquad \text{(XIII-6)}$$

Thus, we can say that the coenzyme has been *oxidized/reduced*.

reduced

**17** Because the coenzyme NAD is reduced, the carbon chain of glucose, in being converted to pyruvic acid, must be ________________.

oxidized

**18** The oxidization of food materials is an *exergonic* process; thus we can conclude that the process of glycolysis results in the release of ________________.

energy

**19** About 40% of the energy released is *captured* and stored for future use. This concept brings us to the third process that is part of glycolysis, *ATP formation*.

**20** *Energy capture* takes place by the process shown in the equation:

$$\text{Energy} + ADP + P_i \underset{\text{(exergonic)}}{\overset{\text{(endergonic)}}{\rightleftharpoons}} ATP \qquad \text{(XIII-7)}$$

As described earlier, ATP stands for the nucleotide ________________ ________________ (Table 7). ADP indicates the same basic

adenosine triphosphate

phosphates molecule, but with only 2 ________________, and $P_i$ is simple in-
phosphate organic ________________.

**21** This equation shows a reversible reaction. The left-to-right equation, which is *endergonic,* represents an organism's major method of
storage energy capture and ________________.

**22** The reverse reaction (right to left) is the major method for energy
exergonic supply or release. It is strongly *endergonic/exergonic.*

**23** The general importance of equation XIII-7 cannot be overstressed. Here we first encounter ATP, a molecule that plays a key role in almost every aspect of metabolism.

nucleotide **24** To summarize, ATP is chemically a ________________. It is a *high-energy compound* critically involved in the metabolic capture,
energy storage, and release of ________________.

stored **25** When $P_i$ is enzymatically added to ADP, energy is *stored/released.*

**26** When ATP is enzymatically degraded to ADP and $P_i$, energy is
released ________________.

**27** From equation XIII-4, we can summarize the three processes that occur during glycolysis: glucose P is oxidized (through dehydrogena-
pyruvic acid tion) to ________________ ________________; the hydrogen atoms
NAD removed are transferred to the coenzyme ________; and the energy
ADP; $P_i$ released is used to synthesize ATP from ________ and ________.

**28** Remember that it requires nine sequential reactions to do this and
enzyme that each reaction is catalyzed by its own specific ________________. These reactions are grouped together as the process called
glycolysis ________________.

**29** These first steps in cellular respiration have been studied at length, not because they are the most important but because they introduce
ATP us to the energy storage molecule ________ and to the hydrogen trans-
NAD fer coenzymes, of which ________ is an example.

**30** Let us consider next what happens to pyruvic acid as respiration progresses. Two options are available, depending on the presence or absence of oxygen. If a cell has no oxygen available to it, *anaerobic* conditions are said to prevail. Under these conditions, in muscle and some bacteria, pyruvic acid is converted to *lactic acid,* a waste product. Yeast cells convert pyruvic acid to alcohol and $CO_2$. These anaerobic methods of disposing of pyruvic acid following glycolysis are called *fermentations.* Fermentation occurs only if
oxygen there is no ________________ available to the cell.

**31** Fermentation is not oxidative; therefore, by itself the process does
ATP not provide any energy, and no ________ is synthesized.

**32** Pyruvic acid can be metabolized anaerobically in the process called
fermentation ________________.

**33** In most tissues under normal conditions, oxygen is in adequate supply, and *aerobic* conditions are said to prevail. Under these conditions, pyruvic acid is further catabolized, with the release of
energy much more ________________.

**34** The release of energy from pyruvic acid constitutes the next stage of cellular respiration, the first step of which is called *oxidative decarboxylation:*

$$\text{pyruvic acid} + \text{NAD} + \text{CoA} \longrightarrow \text{acetyl-CoA} + \text{NADH}_2 + \text{CO}_2 \qquad \text{(XIII-8)}$$

pantothenic acid

CoA is an abbreviation for coenzyme A, which contains the B vitamin ________________ ____________.

acetyl

**35** Acetyl CoA is an ester of acetic acid and the coenzyme. Thus, coenzyme A can be considered a carrier of ________________ radicals.

reduced ($NADH_2$)

**36** Note that again NAD accepts hydrogen atoms and becomes ________________.

reduction

**37** Acetyl CoA next enters a cyclical metabolic pathway called the *citric acid cycle* (Krebs cycle). In the many steps that follow, the remainder of the glucose molecule, the acetyl group, is degraded. Carbon dioxide is released, and for each acetyl group 8 hydrogen atoms (4 pairs) are also released. Thus we can predict the subsequent ________________ of coenzyme carriers.

aerobic

**38** The citric acid cycle involves the degradation of pyruvic acid under ________________ conditions.

glycolysis
fermentation
glycolysis
oxidative dicarboxylation
citric acid cycle

**39** We can now outline the catabolism of glucose by the following steps:

Anaerobic respiration 1. ________________
2. ________________

Aerobic respiration 1. ________________
2. ________________
3. ________________

electron

**40** The terminal stage in aerobic cellular respiration is not concerned with the carbon atoms of the sugar, but with the hydrogen atoms and *electron transport.* A hydrogen atom was described earlier as composed of a proton ($H^+$) and a(n) ________________.

reduced

**41** The electron transport system is a linked sequence of carrier coenzymes, each accepting hydrogen electrons from one neighboring carrier and passing them on to another (Fig. 5). Thus, during this process sequential carriers are alternately ________________ and oxidized.

exergonic

**42** At certain points along the electron transport pathway, oxidation of a carrier releases a significant amount of energy. This energy is used to manufacture ATP. Electron transport is an ________________ process.

electron transport

**43** Under normal physiological conditions, each pair of hydrogen atoms yields enough energy to make 3 molecules of ATP when passed along the ________________ ________________ system.

**44** At the end of the electron transport system, the enzyme *cytochrome oxidase* catalyzes the reaction:

$$2\,\text{H} + \tfrac{1}{2}\,\text{O}_2 \longrightarrow \text{H}_2\text{O} \qquad \text{(XIII-9)}$$

$H_2O$

Thus, under aerobic conditions all hydrogens of a glucose molecule ultimately end up in the compound ______.

**45** At each point indicated by an asterisk in Fig. 6, 2 hydrogens are

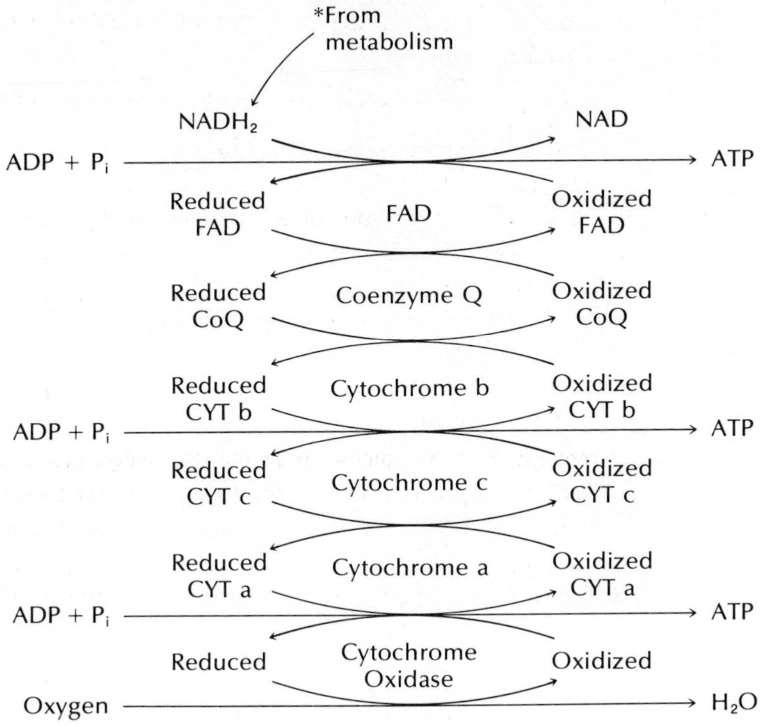


**Fig. 5.** Electron transport. Hydrogen electrons (starting at $NADH_2$) are passed down through the coenzyme carriers and at the cytochrome oxidase level combine with hydrogen ions (formed by the removal of electrons at $NADH_2$ level) and oxygen to yield water. Asterisk indicates that $NADH_2$ is product of citric acid cycle.

transferred to the electron transport system. Remembering that each glucose yields 2 molecules of pyruvic acid, we see that

| | | |
|---|---|---|
| 1 glucose | ⟶ | 2 ATP directly (net) |
| 1 glucose | ⟶ | 12 pairs of hydrogen atoms (2H) |
| each 2 H | ⟶ | 3 ATP in electron transport |

Thus, the energy contained in 1 glucose molecule is transferred to

38 (3 × 12 + 2) ________________ ATP molecules.

**46** Calorimetry shows that as much as 45% of the glucose energy is successfully transferred to ATP; the rest is lost as heat.

**47** We get some use out of the lost heat energy, however, because it keeps our bodies warm. In warm-blooded animals, one purpose of

temperature cellular respiration is the maintenance of body ____________.

**48** We can summarize the complex vital process of cellular respiration in one simple equation:

$$C_6H_{12}O_6 + 6\ O_2 \longrightarrow 6\ CO_2 + 6\ H_2O + \text{energy} \qquad \text{(XIII-10)}$$

This process takes place in every living cell of every plant and animal

respiration species yet studied. Cellular ________________ is a universal catabolic process.

**49** Let us now look briefly at carbohydrate *anabolism*. These processes are not reversals of the catabolic pathways; they involve completely

enzymes different steps and ________________.

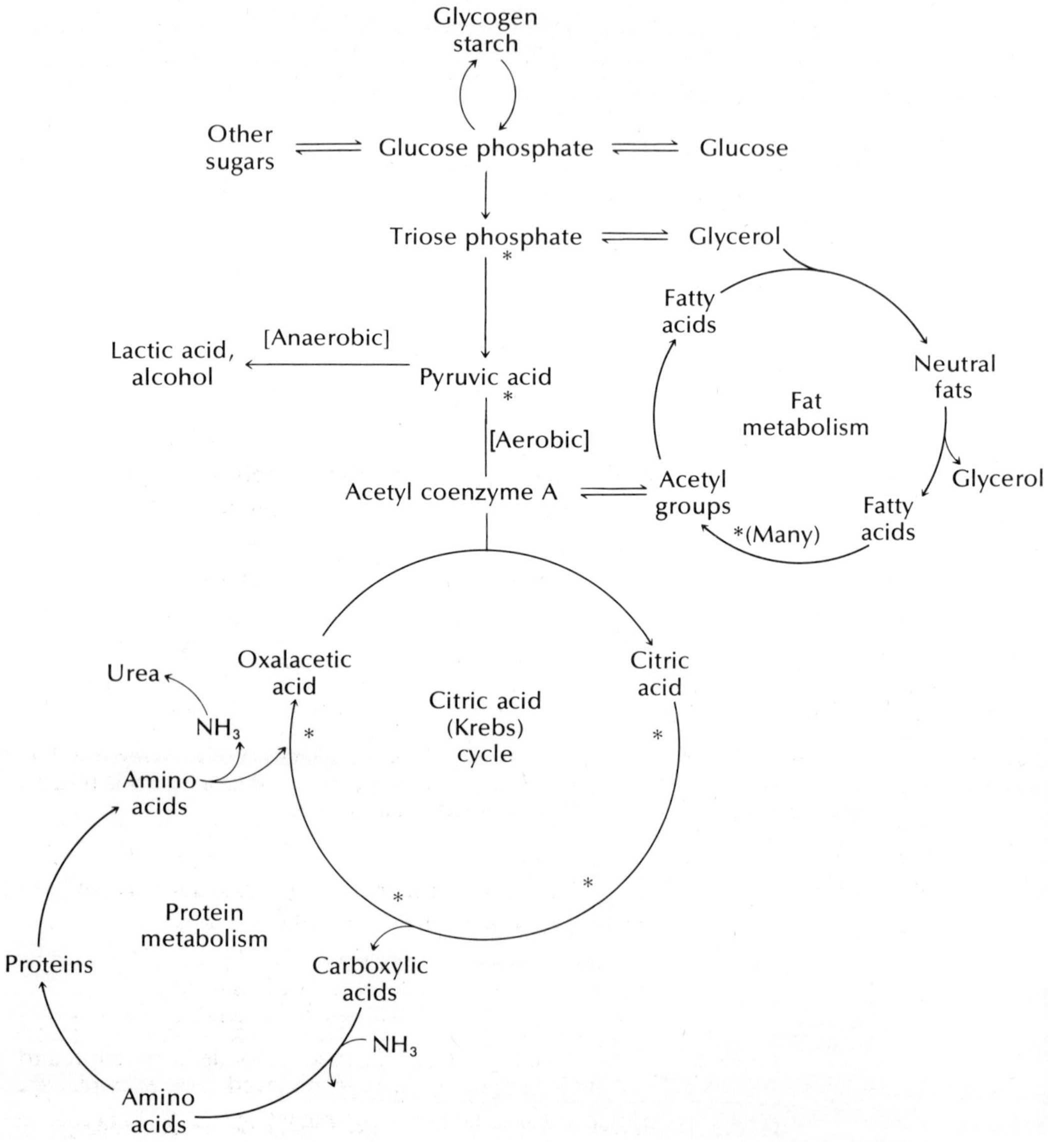


**Fig. 6.** An overview of metabolism. Asterisks show points at which pairs of hydrogens are transferred to the electron transport system.

**50** The most important anabolic process takes place only in green plants and is called photosynthesis. We can write the summary equation as:

$$6\,CO_2 + 6\,H_2O + \text{solar energy} \longrightarrow C_6H_{12}O_6 + 6\,O_2 \qquad \text{(XIII-11)}$$

sun

**51** In photosynthesis, $CO_2$ is *reduced* to yield sugars. This reduction is an endergonic reaction, and the energy is supplied by the ________.

respiration

**52** The sugars and oxygen produced in photosynthesis serve as the raw materials for cellular ________________.

photosynthesis

**53** Thus, plants produce food and oxygen for us and for themselves through ________________.

**54** In animals, an important anabolic reaction is the synthesis of the polysaccharide *glycogen* (animal starch) from glucose phosphate in the liver and muscles. If more glucose is ingested in the diet than can be utilized in respiration, the excess can be stored as ______________ for later use.

glycogen

**55** Glycogen serves as reserve chemical energy; that is, in times of energy need, glycogen is *depolymerized* to yield the monomer ________________-phosphate.

glucose

**56** The building up, or synthesis, of glycogen is called *glycogenesis;* the breaking down, or ________________, of glycogen is glycogenolysis.

depolymerization

## Section 5: LIPID METABOLISM

**1** The study of the metabolism of lipids is complicated by the fact that most natural fats are poorly defined *mixtures* of lipid compounds. Most dietary and human body lipids are triacylglycerols (fats and oils), which are esters of ________________ and ________________ ____________.

glycerol; fatty acids

**2** Let us look first at lipid catabolism. We ingest varying amounts of fat in our diets. Fats, like all lipids, *are/are not* readily soluble in water and body fluids.

are not

**3** Digestive enzymes are water soluble. To establish contact between lipid and *lipase,* the enzyme that digests fats, an *emulisfying agent* is needed. An *emulsion* is a colloidal mixture of a ____________ in a ________________.

liquid
liquid

**4** *Bile salts* (secreted by the liver in bile) serve as emulsifying ________________ in the small intestine.

agents

**5** The enzyme lipase, once in contact with its substrate the triacylglycerol molecule, catalyzes the following reaction:

$$\text{triacylglycerol} + H_2O \longrightarrow \text{glycerol} + \text{fatty acids} \qquad \text{(XIII-12)}$$

One, 2, or all 3 fatty acids may be removed. For example, the complete hydrolysis of glyceryl tripalmitate yields 1 molecule of ________________ and 3 molecules of ________________ ______.

glycerol; palmitic acid

**6** Fatty acids leave the intestine by way of the circulatory system and are carried to all tissues of the body for further catabolism.

**7** All important fatty acids have an *even* number of carbon atoms. For example, lauric acid has ________ carbon atoms, stearic acid has ________ (see Part XII, Section 2).

12
18

**8** Catabolism of fatty acids is carried out by stepwise removal of *2* carbons at a time. The following is a much simplified equation:

$$H_3\text{—}(CH_2)_{n+2}\text{—}\overset{\overset{\displaystyle O}{\|}}{C}\text{—}OH + CoA + 2\ NAD \longrightarrow CH_3\text{—}(CH_2)_n\text{—}\overset{\overset{\displaystyle O}{\|}}{C}\text{—}OH + 2\ NADH_2 + \text{acetyl CoA} \qquad \text{(XIII-13)}$$

If the starting fatty acid had 18 carbons, the ending molecule would have 16 carbons. Note that the 2 removed carbons appear in ________________-coenzyme A.

acetyl

| | | |
|---|---|---|
| citric acid | **9** | We have previously encountered acetyl CoA as a starting material in the ____________ ____________ cycle. |
| $CO_2$ | **10** | All acetyl CoA produced by fatty acid catabolism also goes directly into the citric acid cycle, where it is completely degraded to ________ and pairs of hydrogens. By way of the electron transport system, the hydrogens are converted to the other byproduct of respiration, |
| $H_2O$ | | ________. |
| electron transport | **11** | Note in equation XIII-13 that 2 molecules of NAD are *reduced* to $NADH_2$. The fatty acid is correspondingly *oxidized*. The hydrogen atoms are then also passed through the ________________ ________________ system. |
| 162 | **12** | Each fully catabolized acetyl group can yield as many as 18 molecules of ATP. An 18-carbon fatty acid will yield 9 acetyl groups, and thus a total of ________ molecules of ATP. |
| fatty | **13** | Fats are the most energy-rich of our foods. Gram for gram, they yield many times the energy of carbohydrates. The body *catabolizes* fats when energy is needed beyond the supply afforded by carbohydrates. When more food energy is available than needed, however, the body converts ____________ acids to fats and other lipids for storage. |
| coenzyme | **14** | The biosynthesis of fatty acids proceeds by a stepwise pathway that basically joins carbons together two at a time. Acetyl CoA groups are withdrawn from carbohydrate catabolism for this purpose. The process involves the vitamin *biotin* as a ________________. |
| reduced | **15** | For each 2 carbon atoms joined together, 2 molecules of $NADH_2$ are *oxidized* to NAD. The growing fatty acid chain is correspondingly ________________. |
| requires (consumes) | **16** | An equivalent of 19 molecules of ATP energy is needed for each 2-carbon addition. Once again, whereas catabolism is exergonic and *yields* energy, anabolism is endergonic and ________________ energy. |
| glycerol | **17** | After fatty acids have been synthesized as described, they may combine with ________________ to yield glycerides, or fats. More complex pathways also lead to the formation of phospholipids, waxes, and steroids. |
| lipase | **18** | Review the basic points of lipids. Dietary triacylglycerols are hydrolyzed with the assistance of the enzyme ________________. |
| acetyl CoA | **19** | The resulting fatty acids are catabolized by the stepwise removal of 2-carbon acetyl groups in the form of ________________ ________. |
| citric acid | **20** | These acetyl units are further catabolized by way of the ____________ ____________, or Krebs, cycle. |
| ATP | **21** | Each 2-carbon acetyl group catabolized will yield enough energy to form 18 molecules of ________. |
| carbohydrate | **22** | Biosynthesis, or anabolism, of fatty acids requires acetyl CoA and $NADH_2$, both supplied by ________________ catabolism. |
| energy | **23** | Lipid synthesis stores large amounts of chemical energy. Body fat is our major reserve ________________ supply. |

catabolize

fats (lipids)

**24** Whenever we ingest less food energy than we need, we *catabolize/anabolize* fats.

**25** If we ingest extra food energy (calories), we accumulate ______.

## Section 6: PROTEIN METABOLISM

**1** Protein metabolism encompasses the anabolism and catabolism of proteins and amino acids.

polymerization

amino

**2** Proteins are formed from amino acids by the process of condensation ______________.

**3** Let us deal first with the protein building blocks, the __________ acids.

essential

**4** Some microorganisms and higher plants can manufacture all 20 amino acids by enzymatic processes within their own cells. Other organisms, including all animals, are unable to synthesize 1 or more of the 20. These must be supplied in the diet. Any amino acid that an organism *cannot* make for itself is called an *essential* amino acid. Humans can only synthesize some of the amino acids, so the others must be supplied by diet as ______________ amino acids.

carboxylic; amino

**5** Biosynthesis of an amino acid can be viewed in two steps. First, the carbon backbone is made in the form of a carboxylic acid; then, an amino group is added. Most of the carboxylic acids are obtained directly from the citric acid cycle with some modifications. Thus this cycle, in addition to its functions in carbohydrate and lipid metabolism, also supplies ____________ acids from which __________ acids can be made.

amino

**6** In humans, amino groups ($—NH_2$) can only be obtained from other amino acids. If there is an excess of 1 amino acid, its amino group may be transferred enzymatically to contribute to the manufacture of another needed amino acid. This process is called *transamination*. The enzymes, several of which are clinically important, are *transaminases*. Thus, some dietary amino acids can serve as sources of ______________ groups for the manufacture of other amino acids.

amino

**7** The catabolism of amino acids also falls into two stages. Amino acids are only broken down to any great extent when we consume more protein than we need or when some abnormal condition exists. The first step in amino acid breakdown is *deamination,* or the removal of the __________ group. This occurs in all tissues, but especially in the liver.

**8** The removal of amino groups can be shown as:

(XIII-14)

$$\text{amino acid} + O_2 + H_2O \xrightarrow{\text{amino acid oxidase}} \text{carboxylic acid} + NH_3 + H_2O + [O]$$

ammonia

Note that the process releases the nitrogen in the form of the toxic substance ______________.

**9** In mammals, ammonia is disposed of through a five-step cyclical process that results in the production of the waste product *urea*. This can be summarized as:

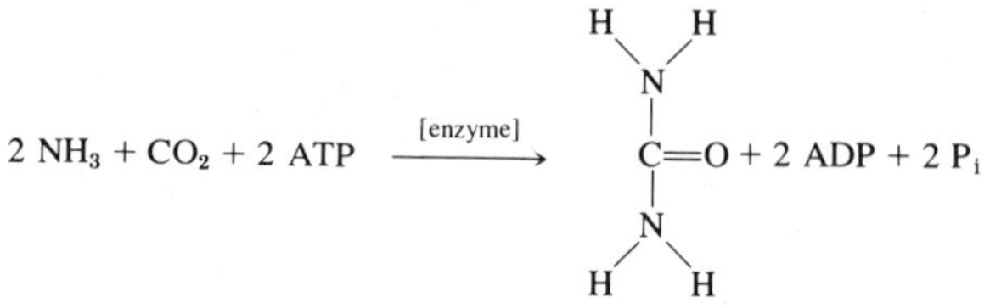


(XIII-15)

**Urea**

In other words, 2 molecules of ammonia plus 1 molecule of carbon dioxide combine to form 1 molecule of ____________.

urea

**10** Clearly, energy is required for urea synthesis, because 2 molecules of _______ are consumed per molecule of urea formed.

ATP

**11** Urea is formed in the liver and excreted by the kidneys in urine. In kidney failure, death may result from buildup of ____________ and other wastes in the blood.

urea

**12** Urinary wastes may be removed artificially by the process of hemo-________________.

dialysis

**13** Let us continue the study of amino acid degradation. After the amino group is removed from an amino acid, the carboxylic acid backbone may enter the citric acid cycle, where it is converted to _______ and water, releasing energy in the process.

$CO_2$

**14** In humans, all 20 amino acids can be degraded to ammonia and carboxylic acids, which are then appropriately disposed of. On the other hand, only 12 amino acids can be synthesized. The other 8 are dietary essentials that must be obtained from the ____________ we eat.

proteins (foods)

**15** We obtain amino acids from dietary proteins by enzymatic hydrolysis. A number of *proteolytic* enzymes, or *proteases,* are secreted in our digestive juices. Each protease takes apart proteins at slightly different points. We can summarize with the equation:

$$\text{protein} + H_2O \xrightarrow{\text{proteases}} \text{amino acids} \qquad \text{(XIII-16)}$$

Because water is involved, this is a(n) ________________ reaction.

hydrolysis

**16** Protein hydrolysis, with the release of amino acids, takes place in the small intestine. The amino acids are then carried by blood circulation to all cells of the body. A well-balanced diet should yield adequate amounts of all _______ *(a number)* amino acids.

20

**17** If one is to remain healthy, the diet *must* yield adequate amounts of the ________________ amino acids.

essential

**18** In the process of digestion, proteins are depolymerized (through hydrolysis) to yield their small subunits (____________ ________), which can be absorbed into the bloodstream.

amino acids

**19** The formation and repair of body tissues hinges on the synthesis of an individual's own specific proteins from absorbed amino acids.

You may recall that this major anabolic event is under *genetic control;* that is, the genes tell a cell *which* amino acids to assemble and in *what order*. A particular gene determines the amino acid

sequence ________________ of a *particular* protein. Thus, 1 gene programs

protein — the assembly of 1 ________________.

**20** The first chemical step in the actual assembly process is called amino acid *activation,* which can be summarized as:

$$\text{amino acid} + \text{ATP} \longrightarrow \text{activated amino acid} \qquad \text{(XIII-17)}$$

ATP — To activate each amino acid, 1 ______ molecule is used.

**21** Activated amino acids are lined up genetically and then joined together by *peptide bonds:*

(XIII-18)

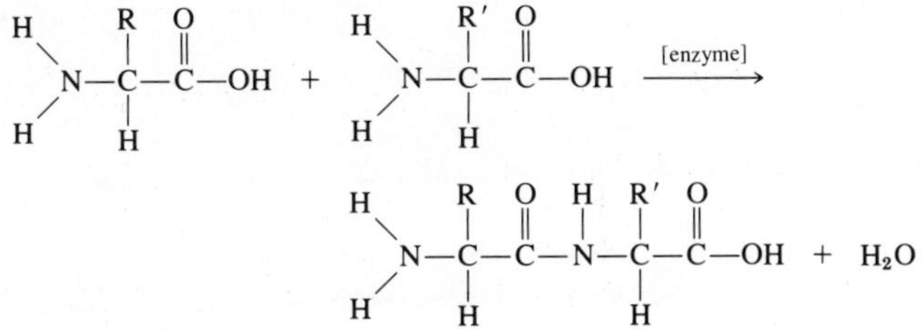


condensation — This is a water-producing ________________ polymerization reaction.

**22** The above reaction goes on, unit by unit, until the protein is assem-

condensation — bled. Thus, a protein is a ________________ polymer.

**23** To assemble any protein, there must be sufficient amounts of the right amino acids; the cell *cannot* substitute one for another. As in making a piece of electronic equipment, if one piece is missing the job cannot be completed. This points out the importance of the essential amino acids: if only *1* is undersupplied, the other amino acids will be wasted and malnutrition will result.

**24** Let us summarize protein metabolism. Proteins ingested in food are

proteases — digested by hydrolytic enzymes, called ________________, in the

intestine — stomach and small ________________.

**25** Free amino acids result, which are distributed to the cells. In cells,

$NH_3$; carboxylic — amino acids may be degraded to yield ______ and ______ acids.

urea; liver — **26** The toxic ammonia is converted to ____________ in the ____________

kidneys — and is then excreted by the ________________.

**27** The carboxylic acids are used for energy production by way of the

citric acid cycle — ____________ ____________ ____________.

**28** In anabolism, carboxylic acids of the citric acid cycle are converted to

amino — amino acids by the addition of an ____________ group.

**29** Amino acids are assembled into proteins by the formation of

peptide — ________________ bonds. This important assembly process is

genetic — under ________________ control.

**30** Because many proteins serve as enzymes, and enzymes control all aspects of metabolism, we can see that all metabolism is ultimately

genes — controlled by the ____________.

## Section 7: NUCLEIC ACID METABOLISM

nucleotides

**1** All foods contain small amounts of *nucleic acids,* which are made up of building blocks called ________________.

pentose

**2** The digestive processes completely degrade nucleic acids into their component nitrogenous bases, phosphate, and ________________ sugars.

RNA

**3** The nitrogenous bases may be further broken down to yield urea and other wastes, or reassembled into specific nucleotides. All normal living organisms are able to manufacture requisite nucleotides and the nucleic acids DNA and ______.

proteins; amino acids

**4** The syntheses of DNA and RNA are key cellular reactions. Both nucleic acids are needed in the transmission and communication of genetic information. In overview, DNA programs the synthesis of *messenger* RNA, which serves as the *template* for the assembly of ________________ from ____________ ____________.

## Section 8: HORMONAL CONTROL OF METABOLISM

**1** *Hormones* are a diverse group of biochemicals that share a common function. In multicellular organisms, hormones serve as *chemical messengers.*

homeostasis

**2** Hormones are formed by one organ and may be conveyed, as by the bloodstream, to another organ, whose function they regulate by means of their chemical activity. In conjunction with the nervous system, hormones regulate and coordinate metabolic processes and thereby maintain balance, or ________________.

amino acids

**3** Chemically, hormones range in complexity from simple amino acids to protein macromolecules, which are condensation polymers of ____________ ____________.

glucose

**4** Generally, a given hormone regulates the function of more than one "target" organ. Epinephrine, for example, speeds up the heart, constricts certain blood vessels, and stimulates the breakdown of glycogen into the monomer ________________-phosphate.

glycogenolysis

**5** Hormones often have opposite, or antagonistic, activities. For example, whereas insulin stimulates glycogenesis, epinephrine stimulates ________________. The relative levels of the two hormones in the blood thus regulate glycogen metabolism in body cells.

epinephrine

**6** In times of physiological stress, we would expect the body to step up its output of *epinephrine/insulin.*

iodine

**7** The hormones with perhaps the most striking effects on total metabolism are insulin and the thyroid hormones *thyroxine* and *triiodothyronine*. Insulin is a relatively simple protein, and thyroxine and triiodothyronine are amino acids whose molecules contain 4 and 3 atoms of ________________ respectively.

**8** Thyroid hormones enhance the activities of all tissues, organs, and systems by stimulating the rate of metabolism. Other things being equal, a *low basal metabolic rate* (BMR) typically points to a deficiency of ________________ hormones. — thyroid

**9** Insulin has a pronounced effect on the metabolism of carbohydrates and fats. Among other metabolic actions, it stimulates the cellular uptake of glucose and enhances the polymerization of ________________ ________________ into glycogen. — glucose phosphate

**10** Based on the functions just cited, it should be clear that insulin *increases/decreases* the blood glucose level. — decreases

**11** The rise or ____________ of a given hormone away from normal levels leads to metabolic derangement and disease. Classic examples are diabetes mellitus, hypothyroidism, and hyperthyroidism. — fall

**12** Diabetes mellitus results from a *lack* of insulin, which explains the *increased* levels of blood glucose associated with the disease. The basic metabolic defect, however, is an impairment of cellular respiration resulting from the lack of *cellular* ________________. — glucose

**13** A lack of thyroid hormones (hypothyroidism) is characterized by a *low/high* BMR; an overproduction (hyperthyroidism) typically produces a *low/high* BMR. — low, high

**14** The signs and symptoms of hypothyroidism and hyperthyroidism clearly relate to a faulty basal ________________ ____________. — metabolic rate

**15** The correction of hormone imbalance and the reestablishment of homeostasis usually requires curtailing the output of the hormone in the event of hypersecretion and administering the hormone in cases of hyposecretion. For example, antithyroid drugs are administered or part of the thyroid gland is removed in ________________, whereas thyroid extract is administered in cases of ________________. — hyperthyroidism, hypothyroidism

# BIBLIOGRAPHY

Alwin, R. H., and Hackworth, R. D.: Algebra—programmed, Books I and II, Englewood Cliffs, N.J., 1969, Prentice-Hall, Inc.

Alwin, R. H., Hackworth, R. D., and Howland, J.: Algebra—programmed, Books III and IV, Englewood Cliffs, N.J., 1971, Prentice-Hall, Inc.

Baker, J. J. W., and Allen, G. E.: Matter, energy and life, ed. 2, Reading, Mass., 1969, Addison-Wesley Publishing Co., Inc.

Brooks, S. M.: Basic science and the human body, St. Louis, 1975, The C. V. Mosby Co.

Christensen, H. N.: pH and dissociation, ed. 2, Philadelphia, 1964, W. B. Saunders Co.

Christensen, H. N.: Neutrality control in the living organism, Philadelphia, 1971, W. B. Saunders Co.

Christensen, H. N., and Palmer, G. A.: Enzyme kinetics, ed. 2, Philadelphia, 1974, W. B. Saunders Co.

Cohen, A.: Handbook of cellular chemistry, St. Louis, 1975, The C. V. Mosby Co.

Drago, R. S.: Principles of chemistry; with practical perspectives, Boston, 1974, Allyn & Bacon, Inc.

White, A., Handler, P., and Smith, E.: Principles of biochemistry, ed. 4, New York, 1968, McGraw-Hill Book Co.